The $RO(G)$-Graded Equivariant Ordinary Homology of G-Cell Complexes with Even-Dimensional Cells for $G = \mathbb{Z}/p$

MEMOIRS
of the
American Mathematical Society

Number 794

The $RO(G)$-Graded Equivariant Ordinary Homology of G-Cell Complexes with Even-Dimensional Cells for $G = \mathbb{Z}/p$

Kevin K. Ferland
L. Gaunce Lewis, Jr.

January 2004 • Volume 167 • Number 794 (fourth of 5 numbers) • ISSN 0065-9266

American Mathematical Society
Providence, Rhode Island

2000 *Mathematics Subject Classification.*
Primary 55M35, 55N91, 55R40, 57S17; Secondary 14M15, 55P91.

Library of Congress Cataloging-in-Publication Data

Ferland, Kevin K., 1969–
The $RO(G)$-graded equivariant ordinary homology of G-cell complexes with even-dimensional cells for $G = \mathbb{Z}/p$ / Kevin K. Ferland, L. Gaunce Lewis, Jr.
p. cm. — (Memoirs of the American Mathematical Society, ISSN 0065-9266 ; no. 794)
'Volume 167, number 794 (fourth of 5 numbers)."
ISBN 0-8218-3461-4 (alk. paper)
1. Homology theory. 2. Fiber spaces (Mathematics) 3. Classifying spaces. 4. Algebraic topology. I. Lewis, L. G. (L. Gaunce), 1949– II. Title. III. Series.

QA3.A57 no. 794
[QA612.3]
510 s—dc22
[514′.23] 2003061937

Memoirs of the American Mathematical Society

This journal is devoted entirely to research in pure and applied mathematics.

Subscription information. The 2004 subscription begins with volume 167 and consists of six mailings, each containing one or more numbers. Subscription prices for 2004 are $583 list, $466 institutional member. A late charge of 10% of the subscription price will be imposed on orders received from nonmembers after January 1 of the subscription year. Subscribers outside the United States and India must pay a postage surcharge of $31; subscribers in India must pay a postage surcharge of $43. Expedited delivery to destinations in North America $35; elsewhere $130. Each number may be ordered separately; *please specify number* when ordering an individual number. For prices and titles of recently released numbers, see the New Publications sections of the *Notices of the American Mathematical Society.*

Back number information. For back issues see the *AMS Catalog of Publications.*

Subscriptions and orders should be addressed to the American Mathematical Society, P. O. Box 845904, Boston, MA 02284-5904, USA. *All orders must be accompanied by payment.* Other correspondence should be addressed to 201 Charles Street, Providence, RI 02904-2294, USA.

Memoirs of the American Mathematical Society is published bimonthly (each volume consisting usually of more than one number) by the American Mathematical Society at 201 Charles Street, Providence, RI 02904-2294, USA. Periodicals postage paid at Providence, RI. Postmaster: Send address changes to Memoirs, American Mathematical Society, 201 Charles Street, Providence, RI 02904-2294, USA.

This publication is indexed in *Science Citation Index®*, *SciSearch®*, *Research Alert®*, *CompuMath Citation Index®*, *Current Contents®/Physical, Chemical & Earth Sciences.*
Printed in the United States of America.

∞ The paper used in this book is acid-free and falls within the guidelines established to ensure permanence and durability.
Visit the AMS home page at http://www.ams.org/

10 9 8 7 6 5 4 3 2 1 09 08 07 06 05 04

Contents

Abstract

It is well known that the homology of a CW-complex with cells only in even dimensions is free. The equivariant analog of this result for G-cell complexes is, however, not obvious, since $RO(G)$-graded homology cannot be computed using cellular chains. We consider $G = \mathbb{Z}/p$ and study G-cell complexes constructed using the unit disks of finite dimensional G-representations as cells. Our main result is that, if X is a G-complex containing only even-dimensional representation cells and satisfying certain finiteness assumptions, then its $RO(G)$-graded equivariant ordinary homology $H_*^G(X; A)$ is free as a graded module over the homology H_* of a point. This extends a result due to the second author about equivariant complex projective spaces with linear $\mathbb{Z}/p$-actions. Our new result applies more generally to equivariant complex Grassmannians with linear $\mathbb{Z}/p$-actions.

Two aspects of our result are particularly striking. The first is that, even though the generators of $H_*^G(X; A)$ are in one-to-one correspondence with the cells of X, the dimension of each generator is not necessarily the same as the dimension of the corresponding cell. This shifting of dimensions seems to be a previously unobserved phenomenon. However, it arises so naturally and ubiquitously in our context that it seems likely that it will reappear elsewhere in equivariant homotopy theory. The second unexpected aspect of our result is that it is not a purely formal consequence of a trivial algebraic lemma. Instead, we must look at the homology of X with several different choices of coefficients and apply the Universal Coefficient Theorem for $RO(G)$-graded equivariant ordinary homology.

In order to employ the Universal Coefficient Theorem, we must introduce the box product of $RO(G)$-graded Mackey functors. We must also compute the $RO(G)$-graded equivariant ordinary homology of a point with an arbitrary Mackey functor as coefficients. This, and some other basic background material on $RO(G)$-graded equivariant ordinary homology, is presented in a separate part at the end of the memoir.

Received by the editor March 2, 2003.

2000 *Mathematics Subject Classification.* Primary 55M35, 55N91, 55R40, 57S17; Secondary 14M15 55P91.

Key words and phrases. Bredon-Illman homology, equivariant ordinary homology, Grassmann Manifolds.

Introduction

If X is a CW complex with cells only in even dimensions, then its integral ordinary homology $H_n(X;\mathbb{Z})$ is a free abelian group in every dimension n. Essentially, the goal of this paper is to prove a precise version (Theorem 2.1) of the following equivariant generalization of this result:

THEOREM. *Let $G = \mathbb{Z}/p$ for some prime p, and let X be a finite G-cell complex having only even-dimensional cells. Then the equivariant ordinary homology $H_*^G X$ of X is free.*

Several issues must be addressed to convert this vague assertion into a precise result. Certainly, the restriction to cyclic groups of prime order in this assertion is rather disappointing. After some background material is presented, the reasons for this constraint, and the potential for weakening it, are discussed in this introduction.

A second issue is what sort of cells are to be used in forming X. Usually, cells of the form $G/H \times D^n$ are used to form G-cell complexes. This choice yields a theorem which is trivial to prove, but turns out to be inapplicable to any interesting G-spaces. An alternative type of G-cell (one which occurs naturally in, for example, equivariant complex flag manifolds) must therefore be introduced before we can state our main result precisely.

The third issue is what sort of equivariant homology is intended here. Tied to that is a fourth question regarding the sense in which $H_*^G X$ is free. The obvious candidates for the homology theory are Borel homology and Bredon-Illman homology. It isn't hard to obtain a version of the theorem above for Borel homology, but our interest is in more sensitive theories than Borel theory.

A simple example illustrates the difficulties which arise in trying to obtain a theorem of the desired sort for Bredon-Illman homology. Let V be a nontrivial complex representation of G. Its one-point compactification S^V is surely the sort of space to which such a freeness theorem ought to apply. Nevertheless, if p is a prime dividing the order of G, then the Bredon-Illman homology $H_n^G(S^V;M)$ of S^V with respect to a coefficient system M contains p-torsion unless M is a very unusual coefficient system — such as one consisting entirely of $\mathbb{Z}[1/p]$-modules. This torsion eliminates the possibility of an interesting freeness result for $\mathbb{Z}$-graded Bredon-Illman homology.

This example also illustrates the problem with using cells of the form $G/H \times D^n$. The space S^V can be constructed using cells of this form, and its $\mathbb{Z}$-graded Bredon-Illman homology can be computed via the chain complex derived from this cell structure. Moreover, it is easy to argue that, if all the cells appearing in this cell structure were even-dimensional, then the Bredon-Illman homology of S^V would be torsion-free. Since this homology is not torsion-free, we know that it is not possible to build even as nice a space as S^V out of only even-dimensional cells of the form

$G/H \times D^n$. Indeed, it seems likely that very few spaces can be constructed using only such even-dimensional cells.

There is a simple explanation for these difficulties with $\mathbb{Z}$-graded Bredon-Illman homology. For any reasonably well-behaved coefficient system M, Bredon-Illman homology with coefficients in M is represented by an equivariant Eilenberg-Mac Lane spectrum in the complete equivariant stable category [**14, 17, 18**]. There is an $RO(G)$-graded equivariant homology theory associated to this spectrum. The equivariant analog of the dimension axiom implies that, for this theory, the homology H_*^M of a point vanishes in dimension n for any nonzero integer n. However, this axiom does not force the vanishing of H_*^M in the dimensions associated to nontrivial virtual G-representations. In fact, in those dimensions, H_*^M is full of torsion at the primes dividing the order of G. The reduced Bredon-Illman homology group $\widetilde{H}_n^G(S^V; M)$ is a part of H_*^M and so reflects this p-torsion.

This explanation for the lack of a good freeness theorem for $\mathbb{Z}$-graded Bredon-Illman homology leads us to both the right homology theory and the right notion of freeness. If the coefficient system M is ring-valued, then its Eilenberg-Mac Lane spectrum is a ring G-spectrum, and the associated homology H_*^M of a point is an $RO(G)$-graded ring. One may then ask if the equivariant homology $H_*^G(X; M)$ of a G-space X is free as a module over H_*^M. Even in the nonequivariant case, this is the sort of freeness one expects when working with a generalized, rather than ordinary, homology theory.

Note that, when M is ring-valued, the suspension axiom implies that the reduced $RO(G)$-graded homology $\widetilde{H}_*^G(S^V; M)$ of S^V is a free module over H_*^M. This suggests that we use the unit disks DV of G-representations V as our cells rather than cells of the form $G/H \times D^n$. For such a cell DV, the appropriate meaning of "even-dimensional" is that, for each subgroup H of G, the fixed point space V^H is an even-dimensional real vector space. Beyond merely providing a freeness result of the desired form, this choice has the advantage that, at least when G is a finite abelian group, many interesting spaces have well-understood cell structures of this sort. Hereafter, the term Rep(G)-cell complex is used for a G-cell complex constructed using the unit disks of G-representations as cells. For nonabelian groups, cells of the form DV may not suffice for building all the spaces we wish to consider. Instead, cells of the form $G/H \times DV$, for a subgroup H of G and a G-representation V, or even cells of the form $G \times_H DW$, for a subgroup H of G and an H-representation W, may be needed. Cells of this sort arise naturally in equivariant Morse theory [**23**] and fit nicely into our approach to proving equivariant freeness results. A G-cell complex constructed from cells of the form $G \times_H DW$ is referred to here as a Rep$^*(G)$-cell complex, with the $*$ intended to suggest the possibility of cells associated to representations of subgroups.

The use of $RO(G)$-graded homology and an alternative type of cell complex leads to one other adjustment in our approach. $RO(G)$-graded homology theories are implicitly Mackey functor valued rather than just abelian group valued. This additional structure plays a critical role in our arguments. Thus, hereafter, we think of equivariant homology as Mackey functor valued. The Burnside ring Mackey functor A plays much the same role in the category of Mackey functors that $\mathbb{Z}$ plays in the category of abelian groups. Thus, A is the generic choice for the coefficients in our ordinary theories. Henceforth, the equivariant ordinary $RO(G)$-graded homology of a point with Burnside ring coefficients is denoted by H_*. The expected superscript G is omitted from this notation because we wish to use H_*^M to denote the equivariant ordinary $RO(G)$-graded homology of a point with some Mackey functor M other than the Burnside ring as the coefficient system.

One difficulty arises immediately in trying to prove a freeness theorem for $RO(G)$-graded equivariant ordinary homology. Unlike $\mathbb{Z}$-graded Bredon-Illman homology, this theory cannot be computed in a straightforward fashion from chain complexes. Thus, the naive algebraic argument used to prove the nonequivariant result must be replaced by an alternative argument. Assume that B is a G-space whose $RO(G)$-graded equivariant ordinary homology is free over H_* with even-dimensional generators. Let Y be a G-space obtained from B by adjoining a single even-dimensional cell DV. If we could show that the homology of Y must also be free, then an inductive argument indicates that any finite Rep(G)-cell complex with only even-dimensional cells has free homology. An obvious tactic for trying to prove the freeness of the homology of Y would be to look at the long exact sequence

$$\cdots \longrightarrow H_\omega^G(B;A) \longrightarrow H_\omega^G(Y;A) \longrightarrow \widetilde{H}_\omega^G(S^V;A) \xrightarrow{\partial_\omega} H_{\omega-1}^G(B;A) \longrightarrow \cdots$$

associated to the cell attachment. This is a long exact sequence of modules over H_*, and the reduced homology $\widetilde{H}_*^G(S^V;A)$ of S^V is a free H_*-module on one generator. Thus, if the boundary map ∂_ω vanished for every ω, then the sequence would split, and $H_*^G(Y;A)$ would be a free H_*-module having one generator for each generator of the homology of B and one additional generator coming from the cell DV. Moreover, the dimensions of these generators would be the obvious ones. This is the approach to an equivariant freeness theorem taken by the second author in [**12**]. There it is shown that, if G is a cyclic group of prime order and X is a Rep$^*(G)$-cell complex having even-dimensional cells which are attached in a suitable order, then the homology $H_*^G(X;A)$ of X is free over H_*. Moreover, it is shown that complex projective spaces with linear G-actions have cell structures of the required sort and so have free homology.

There are two obvious defects in this freeness theorem from [**12**]. The first is that, even for $G = \mathbb{Z}/p$, spaces as simple as the Grassmann manifold of complex 3-planes in a typical 6-dimensional complex G-representation V appear not to have a cell structure satisfying the appropriate dimensional restrictions (see Section 7.2). Thus. there is no reason to expect that the boundary maps in the cell-attaching long exact sequences for such spaces are zero. Hence, the simple approach of [**12**] gives us no freeness result for the homology of such a G-space. Even worse, for groups as small as $\mathbb{Z}/p \times \mathbb{Z}/p$ and $\mathbb{Z}/p^2$, there are linear actions on $\mathbb{C}P^2$ for which all the obvious Rep$^*(G)$-cell structures yield a nonzero boundary map in the long exact sequence associated to attaching the 4-cell to the 2-skeleton. Thus, the approach taken in [**12**] cannot be generalized in a useful way to groups larger than $\mathbb{Z}/p$.

If the modules in our cell-attaching long exact sequences were $\mathbb{Z}$-graded, rather than $RO(G)$-graded, then these nonvanishing results would doom our quest for a more general equivariant freeness result. However, as the first author showed in his thesis [**6**], the additional complexity implicit in the $RO(G)$-grading allows a rather strange thing to happen. At least for the group $\mathbb{Z}/p$, rather than producing torsion in the homology of Y, a nonzero boundary map in the cell-attaching long exact sequence simply forces the generators of the homology of Y to appear in unexpected dimensions. This bit of near magic implies that, if $G = \mathbb{Z}/p$ and X is a finite Rep$^*(G)$-cell complex having only even-dimensional cells, then the homology

$H_*^G(X; A)$ of X with Burnside ring coefficients is free over H_*. There is a one-to-one correspondence between the cells of X and the generators of $H_*^G(X; A)$. However, the dimension-shifting forced by a nonzero boundary map depends so subtly on that map that very little can be said about the dimensions of the generators of $H_*^G(X; A)$. Our impression is that, for most spaces, some completely different line of argument, such as shifting to cohomology and looking at cup products, is needed to determine the dimensions of the generators. This dimension issue is discussed further in Remark 2.9 and illustrated in Section 7.2.

Given this freeness result for finite complexes, it is natural to seek an analogous result for infinite complexes. For the classical nonequivariant freeness result and the main result in [**12**], the transition to infinite complexes is elementary because such a complex X can be described as a colimit of finite complexes whose homologies are direct summands of the homology of X. However, the dimension-shifting in our new freeness result makes extending it to an infinite complex X much trickier. One can, of course, still describe X as a colimit of finite complexes. Unfortunately, the homology of a typical finite subcomplex is no longer a direct summand of the homology of X. Moreover, it is quite easy to construct a diagram of finitely generated free H_*-modules whose colimit is not a free H_*-module. There is no reason to believe that these purely algebraic diagrams cannot be realized as the homology of the diagram of finite subcomplexes from which a G-space X is constructed.

The only way to get around this algebraic difficulty seems to be to impose a condition on X which, in essence, implies that the generator associated to any one cell of X participates in only finitely many dimension shifts. In [**6**], the first author worked with cohomology, rather than homology, and this extra condition took the form of an obvious equivariant analog of a finite-type assumption. The condition imposed here is weaker than that in [**6**] and is best understood by looking at the hypotheses of the freeness theorem in [**12**]. These hypotheses require that, if a cell of the form DW is attached after a cell of the form DV and the dimension of W is greater than the dimension of V, then the dimension of W^G must be at least as large as that of V^G. The extra condition imposed here is that, for each cell DV of X, this dimensional restriction from [**12**] can be broken only finitely many times by cells DW added after DV.

Our freeness result differs from the classical nonequivariant result and the result in [**12**] in that it is not an immediate consequence of a purely algebraic result. In Theorem 3.35 of [**6**], the first author shows that one can take the boundary map $\partial : \widetilde{H}_*^G(S^V; A) \longrightarrow H_{*-1}^G(B; A)$ of a legitimate cell-attaching long exact sequence and construct a long exact sequence

$$\cdots \longrightarrow H_\omega^G(B; A) \longrightarrow D_\omega \longrightarrow \widetilde{H}_\omega^G(S^V; A) \xrightarrow{\partial_\omega} H_{\omega-1}^G(B; A) \longrightarrow \cdots$$

of H_*-modules in which D_* is not a free H_*-module. In order to show that $H_*^G(Y; A)$ is not such a non-free H_*-module, it is necessary to consider the homologies of B, Y, and S^V with coefficients other than the Burnside ring Mackey functor and to examine the long exact homology sequences associated to certain short exact coefficient sequences. The obvious way to obtain the homology of B with respect to some other coefficients would be a Universal Coefficient Theorem. Since no such result existed for $RO(G)$-graded equivariant ordinary homology and cohomology at the time [**6**] was written, ad hoc arguments were used to circumvent the need for this result. These arguments are cumbersome and also very unlikely to be extendible to groups other than $\mathbb{Z}/p$. One of our primary goals in writing this paper was to eliminate the need for such ad hoc arguments. Unfortunately, the Universal Coefficient Theorem for equivariant ordinary cohomology seems inherently less powerful that the corresponding result for nonequivariant ordinary cohomology in that it applies only to finite, rather than finite-type, complexes. This weakness was the primary motivation for our shift from cohomology, which is used in [**6**], to homology. The equivariant Universal Coefficient Theorem for going from homology to cohomology is just as powerful as its nonequivariant analog. Thus, the results in [**6**] can be recovered from our results via that theorem.

One of the particularly attractive aspects of the main freeness theorem in [**6**] is that, since it applies when the cell-attaching boundary maps are nonzero, it is reasonable to hope that this result could be extended to groups other than $\mathbb{Z}/p$. However, the proof given in [**6**] is highly computational and requires a thorough understanding of the multiplicative structure of H_*. It is therefore most unlikely that this argument could be extended to groups more complex than $\mathbb{Z}/p$. A second primary goal in preparing this paper was to replace the arguments in [**6**] with other, more easily extended arguments. With the exception of the argument presented in Section 6.3, our arguments are significantly less computational and require a much less complete understanding of the multiplicative structure of H_*. Unfortunately, in that critical section, we must use the freeness of the homology of complex projective spaces with linear $\mathbb{Z}/p$-actions (proven in [**12**]) to construct a few model long exact sequences. It seems clear that establishing the freeness of the homology of complex projective spaces with linear actions is an unavoidable prerequisite to obtaining a general freeness result like ours for larger groups. Since the cell-attaching maps for these spaces tend to be nonzero for larger groups, this is, for the moment, a serious obstruction.

This paper is divided into two parts. The first part, containing Chapters 1 through 7, presents our freeness results and their proofs. The second part, containing Chapters 8 through 12, supplies background information on $RO(G)$-graded equivariant ordinary homology. That background is needed in Part 1, but, since it is of independent interest, it has been separated out to make it more accessible.

Chapter 1 supplies basic information about Mackey functors, equivariant ordinary homology, and G-cell complexes needed to understand the statement of our freeness theorems. Our freeness theorems for both finite and infinite complexes are stated in Chapter 2. That chapter also contains some examples motivating the somewhat mysterious condition on fixed point dimensions contained in our freeness result for infinite complexes. The proofs of our freeness results are quite long. Chapter 3 provides an overview of the entire argument. However, the real heart of the argument is contained in Chapter 4, which deals with a G-space Y formed by adding a single cell DV to a G-space B with free equivariant homology. Readers interested only in finite complexes, and those wishing to understand the details of the finite case before considering the infinite case, should skip to Chapter 4 after reading Section 3.1. The other two sections of Chapter 3 describe the process of obtaining the result for infinite complexes from that for finite complexes. Essentially, that argument involves nothing more than some rather delicate bookkeeping.

The argument presented in Chapter 4 consists of three key steps. Roughly speaking, the first of these is to identify the generators of $H_*^G(B; A)$ whose dimensions are shifted by the attachment of the new cell. The second is to construct a comparsion map θ from a free H_*-module with generators in the appropriate dimensions into $H_*^G(Y;A)$. The third is to show that this free module fits into a long sequence that can be compared, via the map θ, to the cell-attaching long exact sequence. By comparing the two long exact sequences, we are able to show that θ is an isomorphism. The first of these three steps is handled by Proposition 4.5. The second, which requires the computation of $H_*^G(Y;A)$ in several key dimensions, is described in Proposition 4.12. Proposition 4.14, which also involves homology computations, and Proposition 4.7 complete the third step. The rather lengthy homology computations needed to prove Propositions 4.12 and 4.14 are presented in Chapter 5. It is in these computations that the Universal Coefficient Theorem is invoked. Chapter 6 is devoted to the proof of Proposition 4.7. These four propositions are the core of our entire argument.

The last chapter of Part 1 contains a discussion of complex Grassmann manifolds with linear actions by an abelian group. There we prove the freeness of the equivariant ordinary homology of a complex Grassmann manifold with a linear $\mathbb{Z}/p$-action. That chapter also contains an example illustrating the problems associated with determining the exact dimensions of the homology generators. An analogous freeness result for complex orientable homology theories is proven in [**2**]. Such theories are somewhat less influenced by the complexity of the represention ring $RO(G)$ in that the homology of a G-space X in dimension $\omega \in RO(G)$ depends only on the integer $|\omega|$. Comparing their results with ours offers some insight into the ways in which equivariant theories are influenced by their sensitivity to $RO(G)$.

In order to use the Universal Coefficient Theorem in the proof of our main freeness result, we need some basic background information about $RO(G)$-graded equivariant ordinary homology. The primary purpose of Part 2 is to provide this information. In particular, the first two chapters in this part describe the $RO(\mathbb{Z}/p)$-graded $\mathbb{Z}/p$-equivariant ordinary homology H_*^S of a point with an arbitrary Mackey functor S as coefficients. This information about H_*^S leads to several observations about some curious connections among the $\mathbb{Z}/p$-equivariant Eilenberg-Mac Lane spectra for various Mackey functors (see Corollaries 9.3 and 9.6). Chapter 10 discusses the properties of the category of $RO(G)$-graded Mackey functors for any finite group G. In particular, the box product of $RO(G)$-graded Mackey functors is introduced there. Unfortunately, there is still no published proof of a Universal Coefficient Theorem for $RO(G)$-graded equivariant ordinary homology. The best approach for obtaining this result seems to be via an equivariant generalization of the Universal Coefficient Theorem for E_∞-ring spectra and their E_∞-modules contained in [**4**]. This will be provided in [**16**]. However, this generalization cannot be applied to $RO(G)$-graded equivariant ordinary homology until it is shown that equivariant Eilenberg-Mac Lane spectra have the required E_∞ structures. It is widely acknowledged that the required E_∞ structures exist, at least when the group G is finite. Nevertheless, since there is no published proof of the existence of these structures, Chapter 11 contains a short ad hoc proof of the weak form of the Universal Coefficient Theorem for $\mathbb{Z}/p$-equivariant ordinary homology needed in this paper. The last chapter contains some elementary observations about short exact sequences of Mackey functors for the group $\mathbb{Z}/p$.

The authors are indebted to the referee for some valuable insights and to Mike Mandell and Peter May for a number of helpful conversations about the Universal Coefficient Theorem for equivariant ordinary theories.

Part 1

The Homology of $\mathbb{Z}/p$-Cell Complexes with Even-Dimensional Cells

CHAPTER 1

Preliminaries

1.1. Mackey functors for $\mathbb{Z}/p$

Mackey functors for a finite group were first introduced by Green [**7**]; a more abstract approach was given shortly thereafter by Dress [**3**]. Subsequently, several other approaches have been given, a survey of which can be found in [**22**]. Here we mainly use a slight variant of the approach of Green. This approach is usually refered to as the elementary approach and is particularly convenient for the group $G = \mathbb{Z}/p$. In Section 1 of [**12**], a tutorial is given on Mackey functors for $\mathbb{Z}/p$. Except as indicated below, we adopt the notation used there.

A Mackey functor M for $G = \mathbb{Z}/p$ consists of an abelian group $M(G/G)$, a $\mathbb{Z}[G]$-module $M(G/e)$ and two maps

$$\rho : M(G/G) \longrightarrow M(G/e) \quad \text{and} \quad \tau : M(G/e) \longrightarrow M(G/G).$$

The maps ρ and τ are required to be G-equivariant with respect to the trivial action on $M(G/G)$. Moreover, the composite $\rho \circ \tau$ is required to be the trace of the G-action on $M(G/e)$; that is, $(\rho \circ \tau)(x) = \sum_{g \in G} gx$ for all $x \in M(G/e)$. The maps ρ and τ are called the restriction and transfer, respectively. As in [**12**], M is displayed in a diagram

$$\begin{array}{c} M \\ \\ M(G/G) \\ {\scriptstyle\rho}\big\downarrow \big\uparrow{\scriptstyle\tau} \\ M(G/e) \\ \circlearrowleft \\ {\scriptstyle\theta} \end{array}$$

where θ denotes the G-action. Whenever $M(G/e)$ is a p-fold direct sum C^p of copies of an abelian group C and G acts on $M(G/e)$ by permutations, θ is replaced by the notation *perm*. For $G = \mathbb{Z}/2$, θ is replaced by -1 to indicate an action via multiplication by -1. When the G-action is trivial, θ is omitted from the diagram.

A map f between two Mackey functors M and N consists of two homomorphisms,

$$f_G : M(G/G) \longrightarrow N(G/G) \quad \text{and} \quad f_e : M(G/e) \longrightarrow N(G/e).$$

The homomorphism f_e is required to be a G-map, and the two maps f_G and f_e are required to commute with the restriction and transfer maps in the obvious way. The category $\mathcal{M}$ of Mackey functors is a complete and cocomplete abelian category. Kernels, cokernels, and so forth are defined levelwise.

For easy reference, we recall from [**12**] the particular Mackey functors that are of interest to us. In the following diagrams, C denotes an abelian group, and d

denotes an integer prime to p.

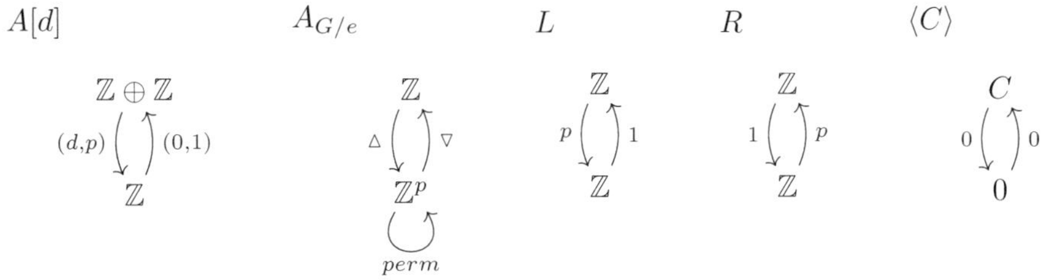

Here, $\triangle$ and ∇ denote the diagonal map and the folding map, respectively. If $G = \mathbb{Z}/2$, then the two additional Mackey functors

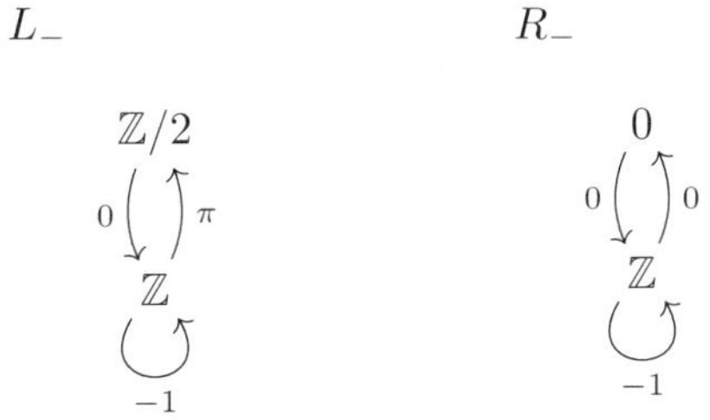

are also of interest. In the display of L_-, π denotes the usual projection. For our computations, it is useful to have standard names for the generators of these Mackey functors. In each of $A[d]$, L, L_-, R, and R_-, ι denotes the generator at G/e. In $A[d]$, L, and L_-, $\tau(\iota)$ is denoted by τ (or by $\bar{\tau}$ if there is a danger that it will be confused with the transfer map τ). Note that τ generates L and L_- at G/G. $A[d]$ is generated at G/G by τ and one other element, which is denoted μ. The significance of the d in the notation $A[d]$ is that $\rho(\mu) = d\iota$. The generator of R at G/G is denoted ξ and is chosen so that $\rho(\xi) = \iota$.

Mackey functors of the form $A[d]$ are projective and play an especially important role in our work. In particular, $A[1]$ is just the Burnside ring Mackey functor A, which plays a role in $\mathcal{M}$ similar to the role played by $\mathbb{Z}$ in the category Ab of abelian groups. It is sometimes useful to employ an alternative set of generators of $A[d]$ at G/G. These alternative generators are denoted σ and κ and are given by

$$\sigma = a\mu + b\tau \qquad \text{and} \qquad \kappa = p\mu - d\tau,$$

where a and b are integers such that $ad + bp = 1$. Note that $\rho(\sigma) = \iota$ and that κ is a generator of the kernel of ρ. The equations

$$\mu = d\sigma + b\kappa \qquad \text{and} \qquad \tau = p\sigma - a\kappa$$

can be used to convert back to the standard basis.

In Examples 1.1(b) of [**12**], the second author shows that $A[d_1] \cong A[d_2]$ if d_1 is congruent to $\pm d_2$ mod p. In fact, the converse is also true.

LEMMA 1.1. *Let d_1 and d_2 be integers prime to p. Then, $A[d_1] \cong A[d_2]$ if and only if $d_1 \equiv \pm d_2 \mod p$.*

PROOF. Our comments prior to the lemma indicate that it suffices to prove that $d_1 \equiv \pm d_2 \bmod p$ if $A[d_1] \cong A[d_2]$. Suppose that $f : A[d_1] \longrightarrow A[d_2]$ is an isomorphism. Let $\{\mu_1, \tau_1, \iota_1\}$ and $\{\mu_2, \tau_2, \iota_2\}$ be standard generators for $A[d_1]$ and $A[d_2]$, respectively. We may assume that $f_e(\iota_1) = \iota_2$, and hence $f_G(\tau_1) = \tau_2$. Write

$f_G(\mu_1) = x\mu_2 + y\tau_2$. The map f_G is described by the 2×2 matrix $\left[\begin{smallmatrix} x & 0 \\ y & 1 \end{smallmatrix}\right]$. Since f_G is an isomorphism, it follows that $x = \pm 1$. The equation

$$d_1\iota_2 = f_e(\rho(\mu_1)) = \rho(f_G(\mu_1)) = (d_2x + py)\iota_2$$

therefore gives the desired congruence. □

The Mackey functor $A_{G/e}$ is a particular instance of a construction due to Dress [**3**]. In general, any Mackey functor M determines a Mackey functor $M_{G/e}$, and the restriction ρ and transfer τ for M determine maps

$$\widehat{\rho}_M : M_{G/e} \longrightarrow M \qquad \text{and} \qquad \widehat{\tau}_M : M \longrightarrow M_{G/e}.$$

The following diagram displays the values of $M_{G/e}$, $\widehat{\rho}_M$, and $\widehat{\tau}_M$.

$$\begin{array}{ccccc}
M_{G/e} & \xrightarrow{\widehat{\tau}_M} & M & \xrightarrow{\widehat{\rho}_M} & M_{G/e} \\
\\
M(G/e) & \xrightarrow{\tau} & M(G/G) & \xrightarrow{\rho} & M(G/e) \\
{\scriptstyle\Delta}\downarrow\uparrow{\scriptstyle\nabla} & & {\scriptstyle\rho}\downarrow\uparrow{\scriptstyle\tau} & & {\scriptstyle\Delta}\downarrow\uparrow{\scriptstyle\nabla} \\
M(G/e)^p & \xrightarrow{\widetilde{\nabla}} & M(G/e) & \xrightarrow{\widetilde{\Delta}} & M(G/e)^p. \\
\circlearrowleft & & \circlearrowleft & & \circlearrowleft \\
perm & & \theta & & perm
\end{array}$$

To define the maps $\widetilde{\nabla}$ and $\widetilde{\Delta}$, we select a generator $g \in G$ and assume that g acts on $M(G/e)^p$ by moving each summand to its successor (mod p). Then, for any $(x_1, x_2, \ldots, x_p) \in M(G/e)^p$ and any $y \in M(G/e)$,

$$\widetilde{\nabla}(x_1, x_2, \ldots, x_p) = \sum_{1 \le k \le p} g^{k-1}x_k \qquad \text{and} \qquad \widetilde{\Delta}(y) = (y, g^{-1}y, \ldots, g^{1-p}y)$$

The maps $\widetilde{\nabla}$ and $\widetilde{\Delta}$ are referred to as the twisted folding and diagonal maps, respectively. Observe that, for each $h \in G$, there is a map $\tilde{h} : M_{G/e} \longrightarrow M_{G/e}$ of Mackey functors which is given by the action of h on $M(G/e)$ at G/G and by a combination of the action of h on each summand and a permutation of the summands at G/e.

The Mackey functors A, $A_{G/e}$, L, and R are each characterized by a universal mapping property. In describing these properties, we denote the abelian group of maps from a Mackey functor M to a Mackey functor N by $\mathcal{M}(M, N)$. Note that evaluation at G/e gives a homomorphism

$$v_{G/e} : \mathcal{M}(M, N) \longrightarrow \mathrm{Hom}_G(M(G/e), N(G/e)).$$

LEMMA 1.2. *Let M be a Mackey functor.*

(a) *The map $\mathcal{M}(A, M) \longrightarrow M(G/G)$ sending a map $f : A \longrightarrow M$ to $f_G(\mu)$ is an isomorphism of abelian groups.*

(b) *Denote by $(1, 0, 0, \ldots, 0)$ the element of $\mathbb{Z}^p = A_{G/e}(G/e)$ which is one in the first coordinate and zero in the others. Then, the map $\mathcal{M}(A_{G/e}, M) \longrightarrow M(G/e)$ sending $f : A_{G/e} \longrightarrow M$ to $f_e((1, 0, 0, \ldots, 0))$ is an isomorphism of abelian groups.*

(c) *The map*

$$v_{G/e} : \mathcal{M}(L, M) \longrightarrow \mathrm{Hom}_G(L(G/e), M(G/e)) = M(G/e)^G$$

is an isomorphism.

(d) *The map*

$$v_{G/e} : \mathcal{M}(M, R) \longrightarrow \mathrm{Hom}_G(M(G/e), R(G/e)) = \mathrm{Hom}(M(G/e)/G, \mathbb{Z})$$

is an isomorphism.

Recall from [**11**] that the category $\mathcal{M}$ carries a symmetric monoidal product which is denoted $\square$ and called the box product. This product plays much the same role in $\mathcal{M}$ as the tensor product plays in *Ab*. Given two Mackey functors M and N, their box product $M \square N$ is described by the diagram

$$M \square N \qquad \begin{array}{c} [(M(G/G) \otimes N(G/G)) \oplus (M(G/e) \otimes N(G/e))]/ \approx \\ (\rho_M \otimes \rho_N, tr) \Big\downarrow \Big\uparrow i_2 \\ M(G/e) \otimes N(G/e) \\ \circlearrowleft \\ \theta_M \otimes \theta_N \end{array}$$

Here, tr denotes the trace map of the action $\theta_M \otimes \theta_N$ of G on $M(G/e) \otimes N(G/e)$. The equivalence relation $\approx$ is determined by the Frobenius relations

$$m_G \otimes \tau n_e \approx \rho m_G \otimes n_e \quad \text{and} \quad \tau m_e \otimes n_G \approx m_e \otimes \rho n_G,$$

where $m_H \in M(G/H)$ and $n_H \in N(G/H)$ for $H = e, G$. The Burnside ring Mackey functor A is the unit for the box product.

In general, box products are difficult to compute, but only a few simple cases are needed in our work. For easy reference, the particular box products of interest to us are recorded in Table 1.1. All of the results displayed there can easily be extracted from Examples 1.2 of [**12**]. Of course, the L_- and R_- entries in this table apply only if $G = \mathbb{Z}/2$.

$\square$	$A[d_2]$	L	R	$\langle D\rangle$	L_-	R_-
$A[d_1]$	$A[d_1d_2]$	L	R	$\langle D\rangle$	L_-	R_-
L	L	L	L	0	L_-	L_-
R	R	L	R	$\langle D/pD\rangle$	L_-	R_-
$\langle C\rangle$	$\langle C\rangle$	0	$\langle C/pC\rangle$	$\langle C\otimes D\rangle$	0	0
L_-	L_-	L_-	L_-	0	L	L
R_-	R_-	L_-	R_-	0	L	L

TABLE 1.1. Box Products

Proposition 1.3 of [**12**] characterizes a map

$$f : M \square N \longrightarrow P$$

out of a box product. The map f determines and is determined by a pair of homomorphisms

$$F_G : M(G/G) \otimes N(G/G) \longrightarrow P(G/G)$$

and

$$F_e : M(G/e) \otimes N(G/e) \longrightarrow P(G/e)$$

which commute with restriction in the obvious way, preserve the G actions at G/e, and respect the Frobenius relations. This characterization is useful for understanding multiplicative structures given by maps out of box products.

The box product can be used to define the notion of a ring in $\mathcal{M}$. Specifically, a Mackey functor ring is a Mackey functor S, together with structure maps

$$\eta : A \longrightarrow S \qquad \text{and} \qquad \phi : S \,\Box\, S \longrightarrow S$$

which specify the unit and multiplication respectively, and which satisfy the usual coherence diagrams. Equivalently, a Mackey functor ring is a Mackey functor S such that ρ is a ring map between the rings $S(G/G)$ and $S(G/e)$, and τ is an $S(G/G)$-module map via ρ. From this, A and R are easily seen to be Mackey functor rings.

For any two Mackey functors M and N, there is a Mackey functor $\langle M, N\rangle$ of "maps" from M to N which provides a right adjoint to the box product construction. This Mackey functor is given by

$$\langle M, N\rangle$$

$$\begin{array}{c} \mathcal{M}(M, N) \\ (\widehat{\rho}_N)_* \Big\downarrow \Big\uparrow (\widehat{\tau}_N)_* \\ \mathcal{M}(M, N_{G/e}). \\ \circlearrowright \\ \theta \end{array}$$

Here, $(\widehat{\rho}_N)_*$ and $(\widehat{\tau}_N)_*$ are derived from the maps $\widehat{\rho}_N$ and $\widehat{\tau}_N$ relating N and $N_{G/e}$, and the action θ comes from the self maps of $N_{G/e}$ associated to the various elements of G. As indicated in Section 1 of [**11**], the adjunction relating $\Box$ and $\langle ?, ?\rangle$ is an isomorphism

$$\mathcal{M}(M \,\Box\, N, P) \cong \mathcal{M}(M, \langle N, P\rangle)$$

natural in each of the three Mackey functors M, N, and P.

1.2. *RO*(*G*)-graded Mackey functor-valued homology

Throughout this paper, we work with $RO(G)$-graded, Mackey functor-valued equivariant ordinary homology theories (see [**17–19**]). For the overview of such theories presented in this section, G can be any finite group. This sort of homology theory is determined by its coefficient system, which is a Mackey functor. For any G-space X, virtual G-representation ω, and Mackey functor M, the Mackey functor-valued homology of X in dimension ω with Mackey functor coefficients M is denoted $\underline{H}^G_\omega(X; M)$. The connection between this notion of equivariant ordinary homology and the older notion introduced by Bredon [**1**] and Illman [**10**] is that the Bredon-Illman homology of X in dimension n with respect to the covariant coefficient system derived from M is just $\underline{H}^G_n(X; M)(G/G)$; that is, the value of $\underline{H}^G_*(X; M)$ associated to the trivial G-representation of dimension n and the orbit G/G.

The equivariant ordinary homology theory associated to M is most easily defined in terms of the equivariant Eilenberg-Mac Lane spectrum HM (see [**14, 15,**

17]). If ω is represented by the formal difference $V - W$ of G-representations V and W and K is a subgroup of G, then the value of the Mackey functor $H_\omega^G(X; M)$ at G/K is given by

$$H_\omega^G(X; M)(G/K) = [\Sigma^V \Sigma^\infty G/K_+, \Sigma^W X_+ \wedge HM]_G,$$

where $[?, ?]_G$ denotes maps in the G-stable category. The restriction and transfer maps for $H_\omega^G(X; M)$ come from the stabilization of space-level maps between orbits and the transfers associated to those space-level maps regarded as equivariant covering spaces (see Corollary V.9.4 and Proposition V.9.9 of [**18**]). At times, we work with the reduced homology $\widetilde{H}_\omega^G(X; M)$ of a based G-space X. This can be viewed either as the homology of the pair $(X, *)$ or as the collection of equivariant stable homotopy groups $[\Sigma^V \Sigma^\infty G/K_+, \Sigma^W X \wedge HM]_G$.

The properties of equivariant ordinary homology used in this paper all follow easily from this spectrum-level definition. In particular, equivariant ordinary homology satisfies the following axioms:

(i) Additivity: Disjoint unions of G-spaces are carried to direct sums.
(ii) Exactness: Cofibre sequences of G-spaces are converted to long exact sequences
(iii) Exactness with respect to coefficients: Any short exact sequence of coefficient Mackey functors yields a long exact sequence in homology.
(iv) Suspension: $\widetilde{H}_\omega^G(X; M) \cong \widetilde{H}_{\omega+V}^G(\Sigma^V X; M)$ for any G-representation V, element ω of $RO(G)$, and based G-space X.
(v) Dimension: For $n \in \mathbb{Z}$, regarded as the trivial G-representation of dimension n,

$$H_n^G(*; M) \cong \begin{cases} M & \text{if } n = 0, \\ 0 & \text{otherwise.} \end{cases}$$

REMARK 1.3. (a) The exactness of homology with respect to coefficients follows directly from the observation that the passage to Eilenberg-Mac Lane spectra converts a short exact sequence of Mackey functors into a fibre sequence of G-spectra.

(b) Note that the dimension axiom says nothing explicit about the homology of a point at the nontrivial elements of $RO(G)$. In fact, as illustrated in Chapter 8, the dimension axiom determines $H_\omega^G(*; M)$ for every $\omega \in RO(G)$, but computing $H_\omega^G(*; M)$ can be highly nontrivial.

One of the ways in which the Burnside ring Mackey functor A plays much the same role in the category of Mackey functors as that played by the integers in the category of abelian groups is that equivariant ordinary homology with Burnside ring coefficients is universal among equivariant ordinary homology theories in the same way that integral homology is universal among nonequivariant ordinary homology theories. In particular, the equivariant ordinary $RO(G)$-graded homology H_* of a point with Burnside ring coeffficients is a ring, and the homology $H_*^G(X; M)$ of any G-space with any coefficients is a (graded) module over H_*. This follows from the fact that HA is a ring spectrum and, for any Mackey functor M, HM is a module spectrum over HA (see Proposition 5.4 of [**15**]).

The complexity of the representation ring $RO(G)$, compared to that of $\mathbb{Z}$, makes it much harder to visualize the homology $H_*^G(X; M)$ of a G-space X than to visualize the homology of a nonequivariant space. For $G = \mathbb{Z}/p$, this difficulty can be reduced somewhat by employing a simple observation about equivariant homology theories. Let $|\omega|$ and $|\omega^G|$ denote the real dimensions of an element ω of $RO(G)$ and its fixed set ω^G, respectively. Assume that ω and ω' are elements of $RO(G)$ such that $|\omega| = |\omega'|$ and $|\omega^G| = |(\omega')^G|$. Then the action map $H_{\omega'-\omega} \Box H^G_\omega(X; M) \longrightarrow H^G_{\omega'}(X; M)$ is an isomorphism by Proposition 8.12. Moreover, $H_{\omega'-\omega} \cong A[d]$ for some integer d prime to p (see Definition 1.4 and Proposition 1.8 below). Thus, $H^G_{\omega'}(X; M)$ is derivable from $H^G_\omega(X; M)$ by a straightforward algebraic process. In fact, frequently $H^G_{\omega'}(X; M)$ and $H^G_\omega(X; M)$ are isomorphic. It follows that one can, essentially, plot $H^G_*(X; M)$ in the plane by assigning the Mackey functor $H^G_\omega(X; M)$ to the point with integer coordinates $(|\omega^G|, |\omega|)$. Strictly speaking, in order to form this plot, one must select a representative ω of each collection of elements of $RO(G)$ having a common pair $(|\omega^G|, |\omega|)$ of dimensions. However, the uncertainty implicit in this selection process is frequently immaterial.

1.3. The homology H_* of a point

Here, we provide some information about the additive and the multiplicative structure of the equivariant ordinary $RO(G)$-graded homology of a point with Burnside ring coefficients, which we denote by H_*. The expected superscript G is omitted from this notation to suppress clutter and so that H^M_* can be used to denote the equivariant ordinary $RO(G)$-graded homology of a point with some other Mackey functor M as the coefficients. As indicated at the end of the previous section, it is almost possible to display H_* by plotting it in the plane. Our first step in describing H_* is introducing the machinery needed to describe the uncertainty in this plot. At the heart of this machinery is a function d out of the subgroup $RO_0(G)$ of $RO(G)$ consisting of those $\omega \in RO(G)$ such that $|\omega| = |\omega^G| = 0$. This function may be regarded as a homomorphism from $RO_0(G)$ to the quotient group $(\mathbb{Z}/p)^\times / \pm 1$ of the multiplicative group $(\mathbb{Z}/p)^\times$ of nonzero elements of $\mathbb{Z}/p$. When so regarded, d is well defined and uniquely determined. Unfortunately, we often need to think of d as a function from $RO_0(G)$ to $\mathbb{Z}$ whose values are integers prime to p. When so regarded, d is not a homomorphism, is not uniquely determined, and is not even obviously well-defined. These problems with d are tied to picking a representive of each element ω of $RO_0(G)$ as a formal difference $V - W$ of two G-representations V and W. For $p = 2$, no difficulties arise in selecting V and W. For an odd prime p, every element of $RO_0(G)$ can be written as a formal difference of complex G-representations. This suffices to ensure that d is at least well-defined as a function into $\mathbb{Z}$.

DEFINITION 1.4. If $p = 2$, then $RO_0(G) = \{0\}$ and $d : RO_0(G) \longrightarrow \mathbb{Z}$ maps 0 to $1 \in \mathbb{Z}$. If p is odd and $\omega \in RO_0(G)$ is nonzero, select nontrivial irreducible complex G-representations $\eta_1, \eta_2, \ldots, \eta_n$ and $\zeta_1, \zeta_2, \ldots, \zeta_n$ such that

$$\omega = \eta_1 + \eta_2 + \ldots + \eta_n - (\zeta_1 + \zeta_2 + \ldots + \zeta_n).$$

For each i, take d_i to be the least positive integer such that the complex power map $z \mapsto z^{d_i}$ is a G-map from the unit circle $S\eta_i \subset \mathbb{C}$ of η_i to $S\zeta_i$. Note that d_i must be prime to p since both η_i and ζ_i are nontrivial G-representations. Let

$$d(\omega) = \prod_i d_i.$$

Also, let $d(0) = 1$. Observe that this gives a well-defined map from $RO_0(G)$ to $\mathbb{Z}$ whose values are integers prime to p. Typically, we denote $d(\omega)$ by d_ω.

It is easy to verify the following key properties of the function d.

LEMMA 1.5. (a) *When regarded as a map into $(\mathbb{Z}/p)^\times/\pm 1$, d is a homomorphism and is independent of the choices made in its definition.*

(b) *Let η be a nontrivial irreducible complex G-representation, and let η^k be its k-fold complex tensor power for some integer k relatively prime to p. Then $\eta - \eta^k \in RO_0(G)$ and $d_{\eta-\eta^k} \equiv \pm k \mod p$. Thus, when regarded as a map into $(\mathbb{Z}/p)^\times/\pm 1$, d is surjective.*

REMARK 1.6. (a) The integer d_ω depends on the choices made in its definition in several ways. First, it depends on the ordering of the η_i and ζ_i. Second, if a nontrivial irreducible representation η is inserted in both of the η_i and ζ_i lists, but at different places in those lists, then d_ω is changed. These two dependencies vanish if d is regarded as a function into $(\mathbb{Z}/p)^\times$. The most serious dependency comes, however, from the identification of complex representations with their conjugates in $RO(G)$. Replacing one of the η_i or ζ_i with its conjugate changes the sign of d_ω in $(\mathbb{Z}/p)^\times$. Passing to the quotient group $(\mathbb{Z}/p)^\times/\pm 1$ eliminates this change. The appearance of $\pm k$, instead of k, in Lemma 1.5(b) is a reflection of this sign problem.

(b) The point of the map d is that, if $\omega \in RO_0(G)$, then $H_\omega \cong A[d_\omega]$. It usually suffices to think of d as a map into $(\mathbb{Z}/p)^\times/\pm 1$, since this value determines the isomorphism class of $A[d_\omega]$. However, in picking generators of either of the standard forms for $A[d_\omega](G/G)$, the integral value of d_ω is implicitly used.

(c) The function d defined here is not the same as that defined in [**12**], because here we are working with homology rather than cohomology. The values of the two functions are multiplicative inverses in $(\mathbb{Z}/p)^\times/\pm 1$.

(d) If $p = 3$, then $(\mathbb{Z}/p)^\times/\pm 1$ is trivial, and the map $d : RO_0(G) \longrightarrow \mathbb{Z}$ can be taken to be the constant map to $1 \in \mathbb{Z}$. However, this masks, rather than eliminates, the sign problems implicit in replacing complex representations by their conjugates.

The additive structure of H_* depends on whether p is even or odd. Thus, we describe that structure in a two part proposition (which is a special case of Theorem 8.1) and illustrate it in two plots. Since H_ω almost always depends only on the pair $(|\omega^G|, |\omega|)$, the plots displayed on page 17 are likely to be more enlightening than the proposition. To understand both the plot and the part of the proposition applicable to odd primes, it is important to remember that, in this case, the integers $|\omega|$ and $|\omega^G|$ associated to any $\omega \in RO(G)$ are either both even or both odd. Because of this, the typical vertical and horizontal spacing between adjacent nonzero values in the plot for odd primes is 2 units. The existence of a real one dimensional sign representation for $\mathbb{Z}/2$ causes the typical vertical spacing between adjacent nonzero values in the plot for $p = 2$ to be one unit. Except on the horizontal axis, the typical horizontal spacing between adjacent nonzero values in that plot is 2 units. In both plots, all zero values of H_* have been suppressed to avoid cluttering the display.

REMARK 1.7. Even for $p = 2$, the display in part (a) of Proposition 1.8 correctly describes H_ω if $|\omega^G|$ and $|\omega|$ are either both even or both odd. This is the part of

H_* that matters for almost every aspect of our arguments. Thus, the best way to follow the remainder of the paper is to focus on the odd prime case.

PROPOSITION 1.8. (a) *Let p be odd and $\omega \in RO(G)$. Then*

$$H_\omega = \begin{cases} A[d_\omega] & \text{if } \omega \in RO_0(G), \\ R & \text{if } |\omega| = 0 \text{ and } |\omega^G| > 0, \\ L & \text{if } |\omega| = 0 \text{ and } |\omega^G| < 0, \\ \langle \mathbb{Z} \rangle & \text{if } |\omega| \neq 0 \text{ and } |\omega^G| = 0, \\ \langle \mathbb{Z}/p \rangle & \text{if } \begin{cases} |\omega| < 0, \ |\omega^G| > 0, & \text{and } |\omega^G| \text{ is even} \\ \quad\text{or} \\ |\omega| > 0, \ |\omega^G| \leq -3, & \text{and } |\omega^G| \text{ is odd,} \end{cases} \\ 0 & \text{otherwise.} \end{cases}$$

(b) *Let $p = 2$ and $\omega \in RO(G)$. Then*

$$H_\omega = \begin{cases} A & \text{if } \omega \in RO_0(G) = \{0\}, \\ R & \text{if } |\omega| = 0, |\omega^G| > 0, \text{ and } |\omega^G| \text{ is even,} \\ R_- & \text{if } |\omega| = 0, |\omega^G| \geq -1, \text{ and } |\omega^G| \text{ is odd,} \\ L & \text{if } |\omega| = 0, |\omega^G| < 0, \text{ and } |\omega^G| \text{ is even,} \\ L_- & \text{if } |\omega| = 0, |\omega^G| \leq -3, \text{ and } |\omega^G| \text{ is odd,} \\ \langle \mathbb{Z} \rangle & \text{if } |\omega| \neq 0 \text{ and } |\omega^G| = 0, \\ \langle \mathbb{Z}/2 \rangle & \text{if } \begin{cases} |\omega| < 0, \ |\omega^G| > 0, & \text{and } |\omega^G| \text{ is even} \\ \quad\text{or} \\ |\omega| > 0, \ |\omega^G| \leq -3, & \text{and } |\omega^G| \text{ is odd,} \end{cases} \\ 0 & \text{otherwise.} \end{cases}$$

Certain basic features of H_* must be understood to follow the proofs of our freeness theorems. These are best understood by looking at the plot for the case of an odd prime. Note first that, by adding trivial representations of various dimensions to any one element ω of $RO(G)$, we move along a line of slope 1 in the plot. Thus, the existence of only one nonzero value on the line $|\omega| = |\omega^G|$ in the plot is a visual representation of the dimension axiom. The empty first and third quadrants of the plot represent other values of H_* that must vanish because of that axiom. The copies of $\langle \mathbb{Z}/p \rangle$ in the second quadrant are the only nonzero values of H_* associated to elements ω of $RO(G)$ for which both $|\omega^G|$ and $|\omega|$ are odd. It is these odd-dimensional elements that create the possibility of nonzero-boundary maps in the cell-attaching long exact sequences for a $\text{Rep}^*(G)$-cell complex with cells only in even dimensions. Note that no nonzero values of H_* occur on the line $|\omega^G| = -1$. This vanishing line in H_* is the foundation for the main freeness theorem in [**12**]. It also plays a critical role in our arguments. Most finite groups larger than $\mathbb{Z}/p$ lack a corresponding vanishing line in the homology of a point. The absence of such a line is one of the primary obstacles to extending our results to those groups.

The essential features of the multiplicative structure of H_*, which are described in Proposition 1.10 below, are also best understood by looking at the plot for odd primes. The copy of R at the point $(2, 0)$ and the copy of $\langle \mathbb{Z} \rangle$ at $(0, -2)$ contain elements ξ_ω and ϵ_ω, respectively, which generate essentially all of H_* other than

$$
\begin{array}{ccccccccc}
 & & & & & |\omega| & & & \\
 & & & & & \wedge & & & \\
 & & & & & \vdots & & & \\
 & \vdots & & \vdots & & \langle \mathbb{Z} \rangle & & & \\
\cdots & \langle \mathbb{Z}/p \rangle & & \langle \mathbb{Z}/p \rangle & & & & & \\
 & & & & & \langle \mathbb{Z} \rangle & & & \\
\cdots & \langle \mathbb{Z}/p \rangle & & \langle \mathbb{Z}/p \rangle & & & & & \\
\cdots & & L & & L & A[d_\omega] & R & R & \cdots > |\omega^G| \\
 & & & & & \langle \mathbb{Z} \rangle & \langle \mathbb{Z}/p \rangle & \langle \mathbb{Z}/p \rangle & \cdots \\
 & & & & & \langle \mathbb{Z} \rangle & \langle \mathbb{Z}/p \rangle & \langle \mathbb{Z}/p \rangle & \cdots \\
 & & & & & \vdots & \vdots & \vdots &
\end{array}
$$

FIGURE 1.1. H_* for p odd

$$
\begin{array}{cccccccccccc}
 & & & & & & |\omega| & & & & & \\
 & & & & & & \wedge & & & & & \\
 & \vdots & & \vdots & & & \vdots & & & & & \\
\cdots & \langle \mathbb{Z}/2 \rangle & & \langle \mathbb{Z}/2 \rangle & & & \langle \mathbb{Z} \rangle & & & & & \\
\cdots & \langle \mathbb{Z}/2 \rangle & & \langle \mathbb{Z}/2 \rangle & & & \langle \mathbb{Z} \rangle & & & & & \\
\cdots & L_- & L & L_- & L & R_- & A & R_- & R & R_- & R & \cdots > |\omega^G| \\
 & & & & & & \langle \mathbb{Z} \rangle & & \langle \mathbb{Z}/2 \rangle & & \langle \mathbb{Z}/2 \rangle & \cdots \\
 & & & & & & \langle \mathbb{Z} \rangle & & \langle \mathbb{Z}/2 \rangle & & \langle \mathbb{Z}/2 \rangle & \cdots \\
 & & & & & & \vdots & & \vdots & & \vdots &
\end{array}
$$

FIGURE 1.2. H_* for $p = 2$

the values which plot to the origin. In particular, the values of H_* on the positive horizontal axis form essentially a polynomial algebra generated by the elements ξ_ω at $(2,0)$. Similarly, the values on the negative vertical axis form essentially a polynomial algebra generated by the elements ϵ_ω at $(0,-2)$. Everything in the fourth quadrant is simply a product of elements on these two half axes. The standard generator τ in the copy of $A[d_\omega]$ at the origin is infinitely divisible by the generators at $(2,0)$. Everything on the negative horizontal axis arises from this divisibility. Similarly, the standard generator κ in the copy of $A[d_\omega]$ at the origin is infinitely divisible by the generators at $(0,-2)$, and everything on the positive vertical axis comes from this divisibility. Every nonzero element in the second quadrant is infinitely divisible by the generators at both $(2,0)$ and $(0,-2)$. Thus, the entire second quadrant can be thought of as coming from the divisibility of the elements in the copy of $\langle\mathbb{Z}/p\rangle$ at $(-3,1)$.

There are, of course, a variety of coordinate systems that could be used to display H_* in the plane. Our choice of $|\omega^G|$ and $|\omega|$ as coordinates may seem strange at first, especially since it causes the $\mathbb{Z}$-graded part of H_* to appear on a diagonal line. However, the central role in the multiplicative structure of H_* played by the elements ξ_ω and ϵ_ω makes it extremely desirable to select a coordinate system in which those elements lie on the axes.

In order to characterize the projective objects in the category of H_*-modules, we need to describe the additive structure of $H_*^G(G/e\,;A)$.

COROLLARY 1.9. *Let ω be an element of $RO(G)$. Then*

$$H_\omega^G(G/e\,;A) = \begin{cases} A_{G/e} & \text{if } |\omega| = 0, \\ 0 & \text{otherwise.} \end{cases}$$

PROOF. This follows immediately from the observation that, for any G space X and any $\omega \in RO(G)$, $H_\omega^G(G/e \times X; A) \cong H_\omega^G(X; A)_{G/e}$. □

In discussing Figure 1.1, we have already given a rough description of the multiplicative structure of H_*. However, to make the details of that structure easily accessible, we now recall from Theorems 4.3 and 4.9 of [**12**] the portion of it needed for our arguments. We need to understand this structure only in those dimensions ω for which $|\omega^G|$ and $|\omega|$ are either both even or both odd. In these dimensions, it does not matter whether or not p is 2.

PROPOSITION 1.10. *There exist elements*

$$\begin{cases} \iota_\omega \in H_\omega(G/e) & \text{for } |\omega| = 0, \\ \mu_\omega, \bar{\tau}_\omega, \kappa_\omega \in H_\omega(G/G) & \text{for } \omega \in RO_0(G), \\ \xi_\omega \in H_\omega(G/G) & \text{for } |\omega| = 0, |\omega^G| > 0, \text{and } |\omega^G| \text{ even}, \\ \bar{\tau}_\omega \in H_\omega(G/G) & \text{for } |\omega| = 0, |\omega^G| < 0, \text{and } |\omega^G| \text{ even}, \\ \epsilon_\omega \in H_\omega(G/G) & \text{for } |\omega| < 0 \text{ and } |\omega^G| = 0, \\ \epsilon_\upsilon^{-1}\kappa_\omega \in H_{\omega-\upsilon}(G/G) & \text{for } \omega \in RO_0(G),\ |\upsilon| < 0, \text{and } |\upsilon^G| = 0, \\ \nu_\omega \in H_\omega(G/G) & \text{for } |\omega| > 0,\ |\omega^G| \le -3, \text{and } |\omega^G| \text{ odd} \end{cases}$$

of H_* which additively generate $H_\omega(G/G)$ (or $H_\omega(G/e)$ as appropriate) in their dimensions. Moreover, these elements satisfy the relations:

(a) $\iota_\omega \iota_{\omega'} = \iota_{\omega+\omega'}$
(b) $\rho(\mu_\omega) = d_\omega \iota_\omega$

(c) $\bar{\tau}_\omega = \tau(\iota_\omega)$ *if* $|\omega^G| \leq 0$
(d) $\kappa_\omega = p\mu_\omega - d_\omega\bar{\tau}_\omega$
(e) $\rho(\xi_\omega) = \iota_\omega$
(f) $\mu_\omega\xi_\upsilon = d_\omega\xi_{\omega+\upsilon}$
(g) $\xi_\omega\xi_\upsilon = \xi_{\omega+\upsilon}$
(h) $\bar{\tau}_\omega\xi_\upsilon = \begin{cases} \bar{\tau}_{\omega+\upsilon} & \text{if } |(\omega+\upsilon)^G| \leq 0 \\ p\xi_{\omega+\upsilon} & \text{if } |(\omega+\upsilon)^G| > 0 \end{cases}$
(i) $\mu_\omega\epsilon_\upsilon = \epsilon_{\omega+\upsilon}$
(j) $\epsilon_\omega\epsilon_\upsilon = \epsilon_{\omega+\upsilon}$
(k) $\epsilon_\omega\xi_\upsilon$ *generates* $H_{\omega+\upsilon}(G/G)$
(l) $\epsilon_\omega\xi_\upsilon = d_{\upsilon'-\upsilon}\epsilon_{\omega'}\xi_{\upsilon'}$ *if* $\omega+\upsilon = \omega'+\upsilon'$
(m) $\epsilon_\upsilon^{-1}\kappa_\omega = \epsilon_{\upsilon'}^{-1}\kappa_{\omega'}$ *if* $\omega-\upsilon = \omega'-\upsilon'$
(n) $\mu_{\omega'}(\epsilon_\upsilon^{-1}\kappa_\omega) = \epsilon_\upsilon^{-1}\kappa_{\omega+\omega'}$
(o) $\epsilon_{\upsilon'}(\epsilon_\upsilon^{-1}\kappa_\omega) = \begin{cases} \epsilon_{\upsilon-\upsilon'}^{-1}\kappa_\omega & \text{if } |\upsilon-\upsilon'| < 0 \\ \kappa_{\omega+\upsilon'-\upsilon} & \text{if } |\upsilon-\upsilon'| = 0 \\ p\epsilon_{\omega+\upsilon'-\upsilon} & \text{if } |\upsilon-\upsilon'| > 0 \end{cases}$
(p) $(\epsilon_\upsilon^{-1}\kappa_\omega)(\epsilon_{\upsilon'}^{-1}\kappa_{\omega'}) = p(\epsilon_{\upsilon+\upsilon'}^{-1}\kappa_{\omega+\omega'})$
(q) $\mu_\omega\nu_\upsilon = \nu_{\omega+\upsilon}$
(r) $\epsilon_\omega\nu_\upsilon = \nu_{\omega+\upsilon}$ *if* $|\omega+\upsilon| > 0$
(s) $\xi_\omega\nu_\upsilon = d\nu_{\omega+\upsilon}$ *for some integer d prime to p if* $|(\omega+\upsilon)^G| \leq -3$
(t) $(\epsilon_\upsilon^{-1}\kappa_\omega)\nu_{\upsilon'} = 0$

In the statements of these relations, the subscripts indicating the dimensions of the elements are implicitly assumed to be in the allowed range of dimensions for that type of element.

1.4. Modules over H_*

In this section, we introduce the category H_*-Mod of modules over the $RO(G)$-graded Mackey functor ring H_*. Only those properties of H_*-Mod needed to state our main result and to outline its proof are covered here. Most of what we need is related to the behavior of free H_*-modules. The more sophisticated aspects of the category H_*-Mod, like its symmetric monoidal closed structure, are discussed later in Chapter 10.

An H_*-module M may be described as an $RO(G)$-graded collection M_ω of Mackey functors together with action maps

$$H_\upsilon \,\Box\, M_\omega \longrightarrow M_{\upsilon+\omega},$$

for $\upsilon, \omega \in RO(G)$, which make the obvious diagrams commute. In making use of the module structure on M, we often view these action maps in a slightly different way.

Definition 1.11. Assume that $\upsilon \in RO(G)$ and $x \in H_\upsilon(G/G)$. By Lemma 1.2(a), there is a unique map $\tilde{x} : A \longrightarrow H_\upsilon$ which takes the standard generator μ of $A(G/G)$ to x. For any $\omega \in RO(G)$, the composite

$$M_\omega \cong A \,\Box\, M_\omega \xrightarrow{\tilde{x}\Box 1} H_\upsilon \,\Box\, M_\omega \longrightarrow M_{\upsilon+\omega}$$

is referred to as the multiplication by x map on M.

The appropriate definition of a free module in H_*-Mod is not quite as obvious as one would expect. Thus, our first objective is to assign a precise meaning to that notion. To accomplish that goal, we must first describe a natural set of projective generators for the category H_*-Mod. For any H_*-module M and any $\omega \in RO(G)$, let $\Sigma^\omega M$ denote the H_*-module specified by $(\Sigma^\omega M)_v = M_{v-\omega}$. We refer to such a module as a dimension-shifted copy of M. Since H_* is a graded ring, a set of projective generators for H_*-Mod obviously ought to include dimension-shifted copies of H_*. If we were working with abelian groups rather than Mackey functors, this would suffice. However, a somewhat larger set of generators is needed in the context of Mackey functors. The source of this need can be seen even at the level of ungraded Mackey functors. If S is a Mackey functor ring and C is a module over S, then C need not be a quotient of a direct sum of copies of S because the elements of $C(G/e)$ are not clearly seen by S. These elements can only be seen properly by $S_{G/e}$. Thus, for a typical Mackey functor ring S, the obvious set of projective generators for S-Mod is $\{S, S_{G/e}\}$. By analogy with the category of modules over a graded ring, one can think of $S_{G/e}$ as a copy of S shifted by "dimension" G/e. The analog of $S_{G/e}$ for the category H_*-Mod is the homology $H_*^G(G/e; A)$ of a single free orbit G/e. We denote this H_*-module by $(H_*)_{G/e}$. Of course, we need to include dimension-shifted copies of $(H_*)_{G/e}$ in our set of projective generators of H_*-Mod. However, if $v, \omega \in RO(G)$ and $|v| = |\omega|$, then the obvious G-homeomorphism between the G-spaces $\Sigma^v(G/e)_+$ and $\Sigma^\omega(G/e)_+$ induces an isomorphism between $\Sigma^v(H_*)_{G/e}$ and $\Sigma^\omega(H_*)_{G/e}$. Thus, we include only the modules $\Sigma^m(H_*)_{G/e}$, for $m \in \mathbb{Z}$. The following result suffices to prove that the H_*-modules $\Sigma^\omega H_*$ and $\Sigma^m(H_*)_{G/e}$ together form a set of projective generators for H_*-Mod.

LEMMA 1.12. *Let M be a module over H_* and ω be an element of $RO(G)$.*

(a) *The set of H_*-module maps from $\Sigma^\omega H_*$ to M is isomorphic to the abelian group $(M_\omega)(G/G)$.*

(b) *The set of H_*-module maps from $\Sigma^{|\omega|}(H_*)_{G/e}$ to M is isomorphic to the abelian group $(M_\omega)(G/e)$.*

This lemma implies that any direct sum

$$P = \left(\bigoplus_{i\in\mathcal{I}} \Sigma^{\omega_i} H_*\right) \bigoplus \left(\bigoplus_{j\in\mathcal{J}} \Sigma^{m_j}(H_*)_{G/e}\right)$$

of dimension-shifted copies of H_* and $(H_*)_{G/e}$ is projective as an H_*-module. Such a direct sum is, however, much better behaved than an arbitrary projective module in that Lemma 1.12 provides very precise control over maps out of P. We can think of P as having one "generator" in dimension ω_i for each $i \in \mathcal{I}$ and one "generator" in dimension m_j for each $j \in \mathcal{J}$. An H_*-module map from P to any other H_*-module M is determined by what happens on these "generators". Further, there are no "relations" on P which constrain where these "generators" can be sent. This control over maps is the essential characteristic of a free module over an ordinary ring, and so motivates our definition of a free module.

DEFINITION 1.13. A free module P over the ring H_* is a module isomorphic to a direct sum of the form $(\oplus_{i\in\mathcal{I}}\Sigma^{\omega_i}H_*) \oplus (\oplus_{j\in\mathcal{J}}\Sigma^{m_j}(H_*)_{G/e})$. The individual summands $\Sigma^{\omega_i}H_*$ and $\Sigma^{m_j}(H_*)_{G/e}$ of P should be thought of as the generators of P. It is important to distinguish between these two types of summands. We refer to generators of the form $\Sigma^\omega H_*$ as G/G-generators or generators of type G/G. Analogously, generators of the form $\Sigma^m (H_*)_{G/e}$ are referred to as G/e-generators or generators of type G/e.

Unfortunately, the dimension of a G/G-generator of a free H_*-module P is not as well-defined as one might like because the H_*-modules $\Sigma^\omega H_*$ and $\Sigma^{\omega'} H_*$ can be isomorphic even if $\omega' \neq \omega$ in $RO(G)$. The following result describes this uncertainity in the dimension of a G/G-generator.

LEMMA 1.14. *The H_*-modules $\Sigma^\omega H_*$ and $\Sigma^{\omega'} H_*$ are isomorphic if and only if $\omega' - \omega \in RO_0(G)$ and $d_{\omega'-\omega} \equiv \pm 1 \mod p$.*

PROOF. First assume that $\Sigma^\omega H_* \cong \Sigma^{\omega'} H_*$. Then $A = (\Sigma^{\omega'} H_*)_{\omega'} \cong (\Sigma^\omega H_*)_{\omega'}$. However, Proposition 1.8 indicates that $(\Sigma^\omega H_*)_{\omega'}$ cannot be isomorphic to A unless $\omega' - \omega \in RO_0(G)$. If $\omega' - \omega \in RO_0(G)$, then $(\Sigma^\omega H_*)_{\omega'} = A[d_{\omega'-\omega}]$ by the same proposition, and Lemma 1.1 gives that $d_{\omega'-\omega} \equiv \pm 1 \mod p$.

Now assume that $\omega' - \omega \in RO_0(G)$ and $d_{\omega'-\omega} \equiv \pm 1 \mod p$. Use Lemmas 1.1, 1.2(a), and 1.12(a) to select an H_*-module map $f : \Sigma^\omega H_* \longrightarrow \Sigma^{\omega'} H_*$ such that $f_\omega : (\Sigma^\omega H_*)_\omega \longrightarrow (\Sigma^{\omega'} H_*)_\omega$ is an isomorphism. Let $\upsilon \in RO(G)$, and consider the commuting diagram

$$\begin{array}{ccc} H_{\upsilon-\omega} \,\square\, (\Sigma^\omega H_*)_\omega & \xrightarrow[\cong]{1 \square f_\omega} & H_{\upsilon-\omega} \,\square\, (\Sigma^{\omega'} H_*)_\omega \\ \downarrow & & \downarrow \\ (\Sigma^\omega H_*)_\upsilon & \xrightarrow{\quad f_\upsilon \quad} & (\Sigma^{\omega'} H_*)_\upsilon \end{array}$$

which expresses the fact that f is an H_*-module map. By Proposition 8.12, the vertical maps in this diagram are isomorphisms. Thus, f is an isomorphism. $\square$

REMARK 1.15. (a) There is some uncertainty in the dimensions of the G/G-generators of a free H_*-module beyond that described in Lemma 1.14. There are sequences $\upsilon_1, \upsilon_2, \ldots, \upsilon_n$ and $\omega_1, \omega_2, \ldots, \omega_n$ of elements of $RO(G)$ for which the free H_*-modules $\oplus_i \Sigma^{\upsilon_i} H_*$ and $\oplus_i \Sigma^{\omega_i} H_*$ are isomorphic for reasons having nothing to do with either merely reindexing the lists of generators or the isomorphisms coming from Lemma 1.14. Examples of this sort are described in [**6**] in the case $n = 2$, $p > 3$, $(|\upsilon_i^G|, |\upsilon_i|) = (|\omega_i^G|, |\omega_i|)$ for $i = 1, 2$, and either $|\upsilon_1^G| = |\upsilon_2^G|$ or $|\upsilon_1| = |\upsilon_2|$.

(b) The value of $\Sigma^\omega H_*$ in dimension ω is the Burnside ring Mackey functor A. An H_*-module map $f : \Sigma^\omega H_* \longrightarrow M$ is determined by the image $f_\omega(G/G)(\mu)$ of the canonical element μ of $A(G/G)$ in $M_\omega(G/G)$. Thus, we could think of μ as the generator of $\Sigma^\omega H_*$. Similarly, the value of $\Sigma^m (H_*)_{G/e}$ in dimension m is the Mackey functor A_G. An H_*-module map $f : \Sigma^m (H_*)_{G/e} \longrightarrow M$ is determined by the image $f_m(G/e)((1, 0, 0, \ldots, 0)) \in M_m(G/e)$ of the element $(1, 0, 0, \ldots, 0)$ of $A_G(G/e)$ introduced in Lemma 1.2(b). Hence, we could think of $(1, 0, 0, \ldots, 0)$ as the generator of $\Sigma^m (H_*)_{G/e}$. However, we rarely work at the level of elements. It is usually far more productive to think of a generator of a free H_*-module P as the inclusion $\Sigma^\omega H_* \longrightarrow P$ or $\Sigma^m (H_*)_{G/e} \longrightarrow P$ of the appropriate summand rather than as the image of μ or $(1, 0, 0, \ldots, 0)$ under this inclusion.

The dimensions of the free modules of interest in this paper frequently satisfy two special conditions.

DEFINITION 1.16. (a) An element ω of $RO(G)$ is said to be even-dimensional if both $|\omega|$ and $|\omega^G|$ are even. Note that, if p is odd, then these two integers have to be either both even or both odd. A G/G-generator of a free H_*-module is said to be even-dimensional if its dimension is an even-dimensional element of $RO(G)$. A G/e-generator is said to be even-dimensional if its dimension m is an even integer.

(b) An element ω of $RO(G)$ is said to be space-like if $|\omega| \geq |\omega^G| \geq 0$. A G/G-generator of a free H_*-module is said to be space-like if its dimension is space-like. A G/e-generator is said to be space-like if its dimension m is a nonnegative integer.

Having introduced free H_*-modules, we now turn to an investigation of maps between finitely generated free H_*-modules.

DEFINITION 1.17. (a) Let ω and ω' be elements of $RO(G)$. The set of maps from the free H_*-module $\Sigma^\omega H_*$ to the free H_*-module $\Sigma^{\omega'} H_*$ can be identified with the abelian group $(\Sigma^{\omega'} H_*)_\omega(G/G)$. Unless $|\omega| = |\omega'|$ and $|\omega^G| = |(\omega')^G|$, this group is one of the cyclic groups $\mathbb{Z}$, $\mathbb{Z}/p$, or 0. A map $f : \Sigma^\omega H_* \longrightarrow \Sigma^{\omega'} H_*$ is called a standard shift map if it is a generator of this cyclic group. Clearly, f is a standard shift map if and only if f_ω is onto. Since our interest in standard shift maps comes from the role they play in dimension-shifting long exact sequences, we have not defined the notion of a standard shift map in the case where $|\omega| = |\omega'|$ and $|\omega^G| = |(\omega')^G|$.

(b) Assume that M is a finitely generated free H_*-module with G/G-generators in dimensions $\omega_1, \omega_2, \ldots, \omega_m$, and that N is a finitely generated free H_*-module with G/G-generators in dimensions $\omega'_1, \omega'_2, \ldots, \omega'_n$. A map $f : M \longrightarrow N$ is determined by its components $f_{i,j} : \Sigma^{\omega_i} H_* \longrightarrow \Sigma^{\omega'_j} H_*$. The map f is said to be constructed from standard shift maps if $f_{i,j}$ is a standard shift map for every i and j. Note that, if there is a pair i, j such that $|\omega_i| = |\omega'_j|$ and $|\omega_i^G| = |(\omega'_j)^G|$, then there is no map from M to N constructed from standard shift maps.

REMARK 1.18. This notion of a standard shift map is somewhat different from that introduced in [**6**]. The change is forced by our discussion of more complex dimension-shifting long exact sequences than those discussed in [**6**].

DEFINITION 1.19. Four types of standard shift maps $f : \Sigma^\omega H_* \longrightarrow \Sigma^{\omega'} H_*$ are of special interest to us.

(a) Assume that $|\omega| = |\omega'|$, $|\omega^G| > |(\omega')^G|$, and $\omega - \omega'$ is even-dimensional so that $(\Sigma^{\omega'} H_*)_\omega \cong R$. Then f is a standard shift map if and only if $f_\omega(\mu) = \pm\xi$, where ξ is the standard generator of R. Such an f is called a horizontal shift map.

(b) Assume that $|\omega| < |\omega'|$, $|\omega^G| = |(\omega')^G|$, and $\omega - \omega'$ is even-dimensional so that $(\Sigma^{\omega'} H_*)_\omega \cong \langle\mathbb{Z}\rangle$. Then f is a standard shift map if and only if $f_\omega(\mu) = \pm\epsilon$, where ϵ is the standard generator of $\langle\mathbb{Z}\rangle$. Such an f is called a vertical shift map.

(c) Assume that $|\omega| < |\omega'|$, $|\omega^G| > |(\omega')^G|$, and $\omega - \omega'$ is even-dimensional so that $(\Sigma^{\omega'} H_*)_\omega \cong \langle\mathbb{Z}/p\rangle$. Then f is a standard shift map if and only if it is nonzero. Such an f is called a diagonal shift map.

(d) Assume that $|\omega| - |\omega'|$ is an odd positive integer and $|\omega^G| - |(\omega')^G|$ is an odd integer less than -1. Then $(\Sigma^{\omega'} H_*)_\omega \cong \langle\mathbb{Z}/p\rangle$, and f is a standard shift map if and only if it is nonzero. Such an f is called a boundary shift map because the boundary maps in our dimension-shifting long exact sequences are constructed from maps of this type.

Because of the critical role played by boundary shift maps in our dimension-shifting long exact sequences, it is important to know the dimensions in which they are nonzero.

LEMMA 1.20. *Let ω, ω', and v be elements of $RO(G)$ such that $|\omega| - |\omega'|$ is an odd positive integer and $|\omega^G| - |(\omega')^G|$ is an odd integer less than -1. Also, let $f : \Sigma^\omega H_* \longrightarrow \Sigma^{\omega'} H_*$ be a nonzero H_*-module map. Then*

$$f_v : (\Sigma^\omega H_*)_v \longrightarrow (\Sigma^{\omega'} H_*)_v$$

is nonzero if and only if all of the following hold:

(i) $|v^G| - |\omega^G|$ *is even,*
(ii) $|\omega| \geq |v|$,
(iii) $\begin{cases} |v| \geq |\omega'| & \text{if } p = 2 \text{ or} \\ |v| > |\omega'| & \text{if } p \neq 2, \end{cases}$
(iv) $|\omega^G| \leq |v^G| \leq |(\omega')^G| - 3$.

PROOF. Observe that, for any v, f_v can be computed from f_ω by using the H_*-module structures on $\Sigma^\omega H_*$ and $\Sigma^{\omega'} H_*$. We begin by showing that f_v vanishes unless v satisfies the listed conditions. By examining the plots of $\Sigma^\omega H_*$ and $\Sigma^{\omega'} H_*$, it is easy to see that, for most v not satisfying these conditions, at least one of $(\Sigma^\omega H_*)_v$ or $(\Sigma^{\omega'} H_*)_v$ is zero. The only exceptions to this occur on the three lines given by the equations $|v| = |\omega|$, $|v| = |\omega'|$, and $|v^G| = |\omega^G|$. On the line $|v| = |\omega|$, the exceptions occur when either $|\omega^G| > |v^G|$ or $|v^G| = |(\omega')^G|$. In these cases, f_v must be zero because there are no nonzero maps of the forms $L \longrightarrow \langle \mathbb{Z}/p \rangle$ or $R_- \longrightarrow \langle \mathbb{Z} \rangle$. On the line $|v| = |\omega'|$, exceptions occur only if $|v^G| > |(\omega')^G| - 3$ and $p = 2$. In this case, f_v must be zero because there are no nonzero maps from $\langle \mathbb{Z}/2 \rangle$ to R_-. On the line $|v^G| = |\omega^G|$, the exceptions occur only if $|\omega| < |v|$. Here, Proposition 1.10(t) implies that f_v is zero.

Now assume that v satisfies the listed conditions. If $|\omega| = |v|$, then $H_{v-\omega} \cong R$ is generated at G/G by $\xi_{v-\omega}$. Proposition 1.10(s) therefore implies that f_v is nonzero. If $|\omega| > |v|$, then select $v' \in RO(G)$ such that $|(v')^G| = |v^G|$ and $|v'| = |\omega|$. The map $f_{v'}$ is nonzero by our earlier argument. Moreover, multiplication by $\epsilon_{v-v'} \in H_{v-v'}$ gives a monomorphism from $(\Sigma^{\omega'} H_*)_{v'}$ to $(\Sigma^{\omega'} H_*)_v$ by Lemma 8.7. It follows trivially that f_v is nonzero. □

1.5. $\mathrm{Rep}^*(G)$-cell complexes

Our initial definitions in this section apply to any compact Lie group G. A $\mathrm{Rep}^*(G)$-cell complex X is a G-space X together with an increasing sequence of G-subspaces X^n of X such that X^0 is a disjoint union of orbits, X^{n+1} is formed from X^n by attaching G-cells, $X = \cup_n X^n$, and X has the colimit (or weak) topology derived from the subspaces X^n. The G-cells allowed in the formation of X are of the form $G \times_H DV$, where H is a (closed) subgroup of G and DV is the unit disk of a finite dimensional H-representation V. Such a cell is attached to X^n in the process of forming X^{n+1} via an attaching G-map from $G \times_H SV$, where SV is the unit sphere of V, to X^n. The set of cells attached to X^n to form X^{n+1} is denoted $\mathcal{J}^{n+1}$. Note that no restrictions are imposed on the dimension of the cells attached to X^n in the formation of X^{n+1}. A cell $G \times_H DV$ is said to be even-dimensional if the fixed point subset V^K is even-dimensional over $\mathbb{R}$ for every subgroup K of H.

In the case of interest within this paper, $G = \mathbb{Z}/p$ for some prime p. Since the only subgroups of G are G itself and the trivial group, the only types of cells appearing in a $\mathrm{Rep}^*(G)$-cell complex X are those of the form DV, for some G-representation V, and those of the form $G \times D^m$, for some integer m. In cells of the latter type, G acts trivially on the disk D^m. These two types of cells are represented algebraically by the two types of generators which occur in free H_*-modules.

Cell complexes of this form are of interest because they arise naturally from equivariant Morse theory (see, for example, [**23**]). Further, if G is a finite abelian group, then the usual Schubert cell structure on Grassmannian manifolds generalizes in an obvious way to a $\mathrm{Rep}^*(G)$-cell structure on the Grassmannian manifold $G(V,k)$ of k-planes in some G-representation V (see Chapter 7). Here, the action of G on $G(V,k)$ is the obvious one derived from the action of G on V.

For any $n \geq 0$, there is a cofibre sequence

$$X^n_+ \longrightarrow X^{n+1}_+ \longrightarrow \bigvee_{G\times_H DV \in \mathcal{J}^{n+1}} G_+ \wedge_H S^V$$

associated to the attachment of the cells in $\mathcal{J}^{n+1}$ to X^n. Here, S^V is the one-point compactification of V. The point at infinity in S^V is given trivial G-action and taken as the basepoint of S^V. Associated to this cofibre sequence, we have long exact sequences

$$\cdots \longrightarrow H^G_*(X^n;S) \longrightarrow H^G_*(X^{n+1};S) \longrightarrow \bigoplus_{G\times_H DV \in \mathcal{J}^{n+1}} \widetilde{H}^G_*(G_+ \wedge_H S^V;S) \xrightarrow{\partial} H^G_{*-1}(X^n;S) \longrightarrow \cdots$$

and

$$\cdots \longrightarrow \widetilde{H}^G_*(X^n;S) \longrightarrow \widetilde{H}^G_*(X^{n+1};S) \longrightarrow \bigoplus_{G\times_H DV \in \mathcal{J}^{n+1}} \widetilde{H}^G_*(G_+ \wedge_H S^V;S) \xrightarrow{\partial} \widetilde{H}^G_{*-1}(X^n;S) \longrightarrow \cdots$$

in equivariant ordinary homology with any Mackey functor S as coefficients. We refer to these sequences as the cell-attaching long exact sequences of X.

An analysis of the boundary map ∂ in these long exact sequences lies at the very center of the main argument in this paper. That analysis is tricky enough when only one cell is added to X^n in the formation of X^{n+1}, and can become hopelessly complicated when more than one cell is added. To get around this difficulty, we produce an alternative filtration of X in which cells are added one at a time. Since we have not assumed that X has only countably many cells, this filtration has to be indexed on some ordinal $\mathcal{J}$ which may be larger than the set of natural numbers. The simpliest way to form $\mathcal{J}$ is to well order each of the sets $\mathcal{J}^n$ and then take their union with the ordering which makes each element of $\mathcal{J}^m$ less than any element of $\mathcal{J}^n$ if $m < n$. The result is a well ordered set, and so may be thought of as an ordinal number. Notationally, however, it is convenient to think of $\mathcal{J}$ as an abstract ordinal. Associated to each $\alpha \in \mathcal{J}$, there is a closed G-subspace X_α of X. Each $\alpha \in \mathcal{J}$ has an immediate successor in $\mathcal{J}$ which is denoted $\alpha + 1$. The subspace $X_{\alpha+1}$ is formed from X_α by adding a single cell, which is denoted $G \times_{H_\alpha} DV_\alpha$. If β is a limit point in $\mathcal{J}$, then $X_\beta = \cup_{\alpha<\beta} X_\alpha$ and has the colimit topology from the X_α. The space X itself is $\cup_{\alpha\in\mathcal{J}} X_\alpha$ and has the colimit topology from the X_α.

Associated to the cell attachment used to form the subspace $X_{\alpha+1}$ from X_α, we have the cofibre sequence

$$(X_\alpha)_+ \longrightarrow (X_{\alpha+1})_+ \longrightarrow G_+ \wedge_{H_\alpha} S^{V_\alpha}.$$

From this cofibre sequence, we obtain the homology long exact sequences

$$\cdots \longrightarrow H_*^G(X_\alpha; S) \longrightarrow H_*^G(X_{\alpha+1}; S) \longrightarrow \widetilde{H}_*^G(G_+ \wedge_{H_\alpha} S^{V_\alpha}; S) \xrightarrow{\partial} H_{*-1}^G(X_\alpha; S) \longrightarrow \cdots$$

and

$$\cdots \longrightarrow \widetilde{H}_*^G(X_\alpha; S) \longrightarrow \widetilde{H}_*^G(X_{\alpha+1}; S) \longrightarrow \widetilde{H}_*^G(G_+ \wedge_{H_\alpha} S^{V_\alpha}; S) \xrightarrow{\partial} \widetilde{H}_{*-1}^G(X_\alpha; S) \longrightarrow \cdots.$$

We refer to this new filtration of X as the "one cell at a time" filtration. In order to simplify the proof of our main theorem, we insist that this new $\mathcal{J}$-indexed filtration of X be consistent with our original filtration of X indexed on the nonnegative integers in that, if $\alpha, \beta \in \mathcal{J}$ and the cells $G \times_{H_\alpha} DV_\alpha$ and $G \times_{H_\beta} DV_\beta$ are in $\mathcal{J}^m$ and $\mathcal{J}^n$, respectively, then $\alpha < \beta$ whenever $m < n$. The geometry need not force this consistency since a cell in $\mathcal{J}^n$ may be attached only to cells in a much lower filtration. However, if $\mathcal{J}$ is constructed in the suggested way from the sets $\mathcal{J}^n$, then the desired consistency is automatic.

For the proof of our main theorem, it is also convenient to regard the orbits contained in X^0 as cells which have been attached in the process of forming X. Observe that, if we regard an orbit G/H as having the empty set $\emptyset$ as its boundary, then attaching it to a G-space B via the unique map of $\emptyset$ into B produces the disjoint union $B \cup G/H$. Thus, we may take the subspace X_0 of X associated to the minimal element 0 of $\mathcal{J}$ to be the empty set, and begin the process of forming X from X_0 by adding the orbits in X^0 to X_0. Essentially, this amounts to revising our original definition of a Rep$^*(G)$-cell complex so that we begin our filtration of X with $X^{-1} = \emptyset$, and form X^0 from X^{-1} by adjointing a collection $\mathcal{J}^0$ of orbits.

CHAPTER 2

The main freeness theorem (Theorem 2.6)

Our freeness theorem for arbitrary $\mathrm{Rep}^*(G)$-cell complexes has two rather odd limitations. These are best understood by looking first at the special case of that theorem applicable to finite complexes and then looking at several examples.

THEOREM 2.1. *Let $G = \mathbb{Z}/p$ and let X be a finite $\mathrm{Rep}^*(G)$-cell complex formed from only even-dimensional cells. Then the $RO(G)$-graded Mackey functor-valued equivariant ordinary homology $H_*^G(X; A)$ of X with Burnside ring coefficients is free over H_*. Moreover, there is a one-to-one correspondence between the generators of $H_*^G(X; A)$ and the cells of X.*

This result is weaker than one might expect in that it does not claim that the generator of $H_*^G(X; A)$ associated to a cell DV of X is in dimension V. Typically, the dimension of the generator associated to a cell DV is only vaguely related to V. Moreover, this dimension can be rather hard to determine. The following example illustrates this dimensional misbehavior in the most easily described case.

EXAMPLE 2.2. Let B be a finite $\mathrm{Rep}^*(G)$-cell complex whose reduced homology $\widetilde{H}_*^G(B; A)$ is free over H_* with G/G-generators in even dimensions $\omega_1, \omega_2, \ldots, \omega_n$ such that

$$|\omega_1| < |\omega_2| < \ldots < |\omega_n|$$

and

$$|\omega_1^G| < |\omega_2^G| < \ldots < |\omega_n^G|.$$

Also, let X be a G-space formed from B by adding an even-dimensional cell DV such that $|V| > |\omega_n|$ and $|V^G| < |\omega_1^G|$. In this case, it is possible for the boundary map

$$\partial : \widetilde{H}_*^G(S^V; A) \longrightarrow \widetilde{H}_{*-1}^G(B; A)$$

in the cell-attaching long exact sequence to hit each of the generators of $\widetilde{H}_*^G(B; A)$ in the sense that its composite with the projection onto the summand spanned by any one generator is nonzero. If this occurs, then $\widetilde{H}_*^G(X; A)$ is free over H_* with G/G-generators in dimensions $\omega_1', \omega_2', \ldots, \omega_{n+1}'$ such that

$$|\omega_i'| = |\omega_i|, \text{ for } i \leq n;$$
$$|\omega_{n+1}'| = |V|;$$
$$|(\omega_i')^G| = |\omega_{i-1}^G|, \text{ for } i \geq 2;$$

and

$$|(\omega_1')^G| = |V^G|.$$

The relations among these various dimensions are best understood via the plot in Figure 2.1. Note that, in this case, none of the generators of $\widetilde{H}_*^G(X; A)$ are in

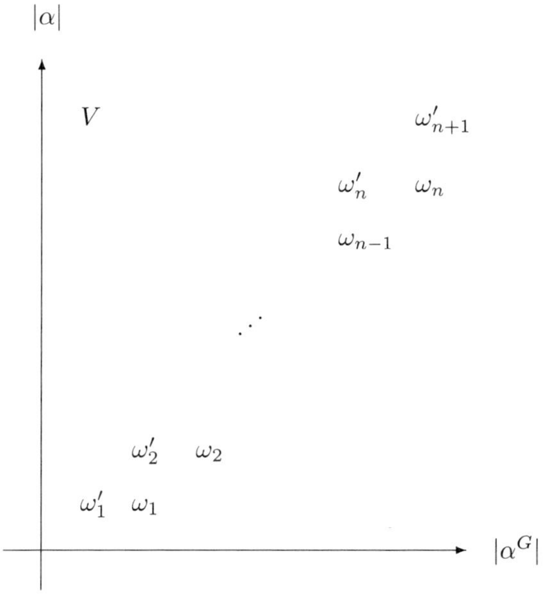

FIGURE 2.1. The generators of $\widetilde{H}^G_*(B;A)$ and $\widetilde{H}^G_*(X;A)$

the expected dimensions. The first n generators of $\widetilde{H}^G_*(X;A)$ are, in a suitable sense, derived from the generators of $\widetilde{H}^G_*(B;A)$. However, their dimensions plot to the left of dimensions of the corresponding generators of B. The last generator is derived from the cell added to B to form X, but its dimension plots to the right of where it might be expected to lie.

In Remark 1.15(a), we noted that two free modules over H_* might be isomorphic even if their generators were in not in the same dimensions. In order to prevent that remark from causing some confusion here, it is important to note that $\widetilde{H}^G_*(X;A)$ is obviously not isomorphic to the free H_*-module with generators in dimensions ω_1, ω_2, $\ldots$, ω_n, and V. Thus, this dimension-shifting is real and not merely a failure to notice that two free H_*-modules with generators in different dimensions are nevertheless isomorphic.

Figure 2.1 and the relations among the dimensions of the elements V, ω_i, and ω'_j of $RO(G)$ represented by that figure play a prominent role in the rest of this memoir. Thus, we adopt some terminology to describe this situation.

DEFINITION 2.3. A sequence ω_1, ω_2, $\ldots$, ω_n of elements of $RO(G)$ satisfying the conditions

$$|\omega_1| < |\omega_2| < \ldots < |\omega_n|$$

and

$$|\omega_1^G| < |\omega_2^G| < \ldots < |\omega_n^G|.$$

is called a ramp of length n. An element V of $RO(G)$ such that $|V| > |\omega_n|$ and $|V^G| < |\omega_1^G|$ is said to bound the ramp. A ramp ω'_1, ω'_2, $\ldots$, ω'_{n+1} of length $n+1$

is said to be a V-shifted ramp associated to the original ramp if the conditions

$$|\omega_i'| = |\omega_i|, \text{ for } i \le n;$$
$$|\omega_{n+1}'| = |V|;$$
$$|(\omega_i')^G| = |\omega_{i-1}^G|, \text{ for } i \ge 2;$$

and

$$|(\omega_1')^G| = |V^G|$$

are satisfied. Note that, if $p \geq 5$, then there is not a unique V-shifted ramp associated to the original ramp because our constraints determine only the integers $|\omega_i'|$ and $|(\omega_i')^G|$ rather than the actual elements ω_i' of $RO(G)$. The combination of a ramp and an associated V-shifted ramp is referred to as a staircase of length n bounded by V. We always assume that the elements in a ramp, and any element said to bound a ramp, are even-dimensional and space-like.

Whenever an even-dimensional cell DV is added to a space B whose homology is free over H_* with even-dimensional generators and the boundary map

$$\partial : \widetilde{H}_*^G(S^V; A) \longrightarrow H_{*-1}^G(B; A)$$

in the cell-attaching long exact sequence is nonzero, some shifting of the dimensions of the generators occurs. However, if the dimensions of the G/G-generators of $H_*^G(B; A)$ hit by the boundary map do not form a ramp like that in Example 2.2, it can be difficult to predict exactly which shifts occur. Hence nothing is said in Theorem 2.1 about the dimensions of the generators of $H_*^G(X; A)$. All that can be said in general about this shifting is that old generators coming from $H_*^G(B; A)$ may remain in the same dimension or move to the left in dimension; whereas the new generator coming from S^V must move to the right of its expected dimension whenever ∂ is nonzero.

This shifting of dimensions means that, even if every finite complex of an infinite $\text{Rep}^*(G)$-cell complex X has homology which is free over H_* with even-dimensional generators, it is not at all obvious that the homology of X must be free over H_*. The following example illustrates what can go wrong.

EXAMPLE 2.4. Let $G = \mathbb{Z}/p$, and let η be a nontrivial irreducible complex G-representation. We want to form a $\text{Rep}^*(G)$-cell complex X containing one 2-cell on which G acts trivially and one cell of the form $D(m\eta)$ for each $m \geq 2$. Let X^1 be S^2 with trivial action, and form X^2 from X^1 by attaching a cell of the form $D(2\eta)$ in such a way that the boundary map ∂ of the cell-attaching long exact sequence is nonzero. Examples of linear actions of G on $\mathbb{C}P^4$ having a cell structure of this form can are given in [**12**]. Since $\partial \neq 0$, the generators of $\widetilde{H}_*^G(X^2; A)$ must plot at the coordinates $(0, 2)$ and $(2, 4)$. One might assume that their dimensions were η and $2 + \eta$. However, due to the way in which dimensions shift, the actual dimensions might involve irreducibles other than η. We would like to form X^3 from X^2 by attaching a cell of the form $D(3\eta)$ in such a way that the boundary map ∂ of the cell-attaching long exact sequence is nonzero on the generator of $\widetilde{H}_*^G(X^2; A)$ which plots to $(2, 4)$. It is not obvious that the desired attaching map can be constructed. However, the algebraic machinery presented in Chapter 6 allows us to construct a purely algebraic dimension-shifting long exact sequence which reflects what must happen in homology if the appropriate cell can be attached. It follows that, if

X^3 can be constructed, then it must have homology generators which plot at the coordinates $(0,2)$, $(0,4)$, and $(2,6)$.

In general, X^m should have homology generators that plot at the coordinates $(0,2)$, $(0,4)$, $\ldots$, $(0,2m-2)$, and $(2,2m)$. The next stage X^{m+1} should be formed from X^m by adding a cell of the form $D((m+1)\eta)$ in such a way that the boundary map ∂ is nonzero on the homology generator of X^m which plots at $(2,2m)$. Again, even though it is not obvious that the desired attaching map exists geometrically, the results in Chapter 6 allow us to construct a purely algebraic long exact sequence which reflects what must happen in homology if the appropriate cell can be attached.

This algebra allows us to see that, if the space $X = \cup_m X^m$ exists, then $\widetilde{H}^G_*(X;A)$ contains a free summand with a generator plotting at $(0,2m)$, for all $m \geq 1$, and another, non-free (and even nonprojective) summand that may be thought of as the "ghost" of the generators plotting at $(2,2m)$ in the various $\widetilde{H}^G_*(X^m;A)$. An easily seen part of this non-free summand is a copy of $\langle\mathbb{Z}\rangle$ at location $(2,2m)$ for every integer m. Whether or not the desired attaching maps exist, this example illustrates the way in which the colimit of a diagram of free $\underline{H}_*$-modules can fail to be free. The technical assumption about fixed point dimensions in our main freeness result is needed to circumvent this colimit problem.

There are two obvious ways to attempt to produce the attaching maps needed to convert this algebraic example into a geometric one. The attaching map needed to form X^2 is of the form $f_2 : S(2\eta) \longrightarrow S^2$, and is known to exist. If this map could be extended over the inclusion of $S(2\eta)$ into $S((m+1)\eta)$, then the composite of this extension with the inclusion of $S^2 = X^1$ into X^m would provide the attaching map needed to form X^{m+1} from X^m. Unfortunately, the Segal conjecture implies that at most finitely many such extensions can exist.

A second approach to producing the map needed to form X^{m+1} would be to attempt to kill off all but the top generator of the homology of X^m by attaching representation disks. Killing these generators would produce a map from X^m to the sphere S^ω, where ω is the dimension of the top generator of the homology of X^m. The map f_2 can be suspended to produce a map $g_{m+1} : S((m+1)\eta) \longrightarrow S^\omega$. Any lifting $f_{m+1} : S((m+1)\eta) \longrightarrow X^m$ of this map would provide the desired attaching map. At least in the equivariant stable category, it is possible to kill off the lower dimensional generators using the stable equivariant Hurewicz Theorem [**13**, Theorem 2.1]. The obstructions to lifting g_{m+1} stably to obtain f_{m+1} live in certain equivariant stable stems. If p is odd, these obstructions must vanish if those equivariant stems contain no p-torsion. If the nonequivariant stems contain no p-torsion in dimensions $2(m-1)$ and below, then the equivariant stems relevant to the construction of f_{m+1} are also free of p-torsion. Thus, for an odd prime p, the standard results on the nonexistence of p-torsion in the lower dimensional stable stems suffice to permit the construction, at least stably, of the first few attaching maps. The question of whether the remaining maps can be constructed appears to be a difficult one.

Our third example illustrates another kind of difficulty which can arise in proving that the homology of an infinite complex is free over $\underline{H}_*$. The homology of the space constructed in this example is, in fact, free over $\underline{H}_*$. However, the proof that it is involves an ad hoc argument which does not seem to have any reasonable generalization.

EXAMPLE 2.5. Recall that, in the first stage of the example above, we attached a cell of the form $D(2\eta)$ to a 2-sphere via an attaching map $f : S(2\eta) \longrightarrow S^2$ that is known to exist. Here, we wish to work with the double suspension of f. Since the boundary $S(2+2\eta)$ of $D(2+2\eta)$ is a suspension, rather than mapping it to a single 4-sphere, we can map it to a wedge $S^4 \vee S^4$ of two 4-spheres by taking one copy of the suspension of f going into each of the two 4-spheres. Now consider an infinite wedge $\bigvee_{m\in\mathbb{Z}} S^4$ of 4-spheres indexed on the integers. Form a new space X by adjoining a $\mathbb{Z}$-indexed collection of cells of the form $D(2+2\eta)$ to this infinite wedge. The mth cell should be adjoined to the infinite wedge by attaching it to the spheres indexed on m and $m+1$ via our sum map into $S^4 \vee S^4$.

The equivariant ordinary homology of the wedge of spheres is certainly free over H_*. However, if one uses the standard generating set for this free module coming from the individual copies of S^4, then the dimension shifting which occurs in the passage to the homology of X makes it very hard to see that the homology of X is also free over H_*. Nevertheless, by replacing the standard generating set with one consisting of exactly one of the standard generators plus the sum of the mth and $(m+1)^{\text{st}}$ standard generators, for every integer m, one can use an elementary variant of the proof of our main freeness theorem to show that $\widetilde{H}^G_*(X;A)$ is free over H_*. It has one generator plotting at the point $(4,4)$, a $\mathbb{Z}$-indexed collection of generators plotting at the point $(2,4)$, and a $\mathbb{Z}$-indexed collection of generators plotting at the point $(4,6)$.

Basis changes similar to the one used here are needed in the proof of our general freeness theorem. However, those basis changes, accomplished in Proposition 4.5, are much less precisely tuned to the geometry of the space than the one used for X. The difference is easily seen by noting that our general result, Theorem 2.6, applies to any subcomplex Y of X which contains only finitely many of the cells of the form $D(2+2\eta)$. Looking over the proof of that result to see how it applies to Y, one can see that the finiteness condition imposed on Y allows us to concoct an appropriate change of basis in a rather naive way. In fact, it would not be unfair to say that we find this change of basis by stumbling around in the dark until we trip over it. Certainly this is a far less elegant approach than beginning the argument, as we did for $\widetilde{H}^G_*(X;A)$, by picking a change of basis ideally suited to the geometry of the space. In part, this lack of elegance seems inherent in the "one cell at a time" approach used to prove our main result. However, for an arbitrary $\mathrm{Rep}^*(G)$-cell complex, it is not at all obvious that there are changes of basis as well suited to a freeness argument as the one we suggest here for X.

Examples 2.4 and 2.5 display the two distinct difficulties motivating the somewhat curious assumption about cell dimensions appearing in our main freeness theorem. It might be possible to weaken this assumption, but it seems unlikely that it can be removed entirely. To understand this assumption, recall that $\mathcal{J}^n$ is the set of cells added in the nth stage X^n of the formation of X.

THEOREM 2.6 (Main Freeness Theorem). *Let $G = \mathbb{Z}/p$ and let X be a $\mathrm{Rep}^*(G)$-cell complex containing only even-dimensional cells. Assume that, for each $m \geq 1$ and each cell DV in $\mathcal{J}^m$, there is only a finite number of cells DW in the collections $\mathcal{J}^n$, for $n > m$, such that $|W| > |V|$ and $|W^G| < |V^G|$. Then the $RO(G)$-graded Mackey functor-valued equivariant ordinary homology of X with Burnside ring coefficients is free over H_*. Moreover, there is a one-to-one correspondence between the generators of $H^G_*(X;A)$ and the cells of X.*

The technical dimensional restriction in Theorem 2.6 is not as restrictive as it might seem. In the nonequivariant context, finite type restrictions often appear. There is an obvious generalization of the notion of finite type for a $\mathrm{Rep}^*(G)$-cell complex, and all finite type $\mathrm{Rep}^*(G)$-cell complexes automatically satisfy the dimensional restriction in this theorem.

DEFINITION 2.7. For $G = \mathbb{Z}/p$, a $\mathrm{Rep}^*(G)$-cell complex X is of finite type if, for each non-negative integer m, X has only finitely many cells of the form $G \times D^n$ with $n \leq m$ and only finitely many cells of the form DV with $|V^G| \leq m$.

If the $\mathrm{Rep}^*(G)$-cell complex X is of finite type, then, for any cell of the form DV in X, there are only finitely many cells of X of the form DW with $|W^G| < |V^G|$. Thus, restricted to $\mathrm{Rep}^*(G)$-cell complexes of finite type, Theorem 2.6 becomes:

COROLLARY 2.8 (Finite Type Freeness Result). *Let $G = \mathbb{Z}/p$ and let X be a finite type $\mathrm{Rep}^*(G)$-cell complex containing only even-dimensional cells. Then the $RO(G)$-graded Mackey functor-valued equivariant ordinary homology of X with Burnside ring coefficients is free over H_*. Moreover, there is a one-to-one correspondence between the generators of $H_*^G(X; A)$ and the cells of X.*

In Chapter 7, we show that the complex Grassmann manifold $G(V, k)$ of complex k-dimensional subspaces of a complex G-representation V is a finite type $\mathrm{Rep}^*(G)$-cell complex having cells only in even dimensions. Thus, this corollary can be used to show that $H_*^G(G(V, k); A)$ is free over H_* (see Corollary 7.2).

The lack of precise information about the dimensions of the generators of $H_*^G(X; A)$ in Theorems 2.1 and 2.6 and Corollary 2.8 is, to say the least, quite disconcerting and very hard to accept. The obstacles to obtaining that information are, however, quite serious. In the remarks below, we offer two ways of understanding the nature of these difficulties.

REMARK 2.9. (a) The boundary map $\partial : \widetilde{H}_*^G(S^V; A) \longrightarrow H_{*-1}^G(B; A)$ of the long exact sequence associated to attaching a cell of the form DV to the G-space B can be regarded as posing an extension problem in the category of H_* modules; that is, find an H_*-module D_* and H_*-module maps $\chi : H_*^G(B; A) \longrightarrow D_*$ and $\psi : D_* \longrightarrow \widetilde{H}_*^G(S^V; A)$ such that the sequence

$$\cdots \longrightarrow H_\omega^G(B; A) \xrightarrow{\chi_\omega} D_\omega \xrightarrow{\psi_\omega} \widetilde{H}_\omega^G(S^V; A) \xrightarrow{\partial_\omega} H_{\omega-1}^G(B; A) \longrightarrow \cdots$$

is exact. We observed in the introduction that there exist non-free H_*-modules D_* which solve this extension problem. The crux of the proof of our freeness results is that the Universal Coefficient Theorem for equivariant ordinary homology can be used to show that such non-free solutions cannot be the homology of the space obtained from B by attaching the cell DV. However, Propositions 4.7 and 4.9 imply that it is also possible to construct non-isomorphic free H_*-modules which solve this extension problem. This is most easily seen by taking B to be the sphere S^{ω_1} associated to an even-dimensional G-representation ω_1 such that $|\omega_1| < |V|$ and $|V^G| < |\omega_1^G|$. If we replace the homology $H_*^G(B; A)$ of B by its reduced homology $\widetilde{H}_*^G(B; A)$ in the extension problem above, then Lemma 6.5, which is a very simple special case of Proposition 4.9, can be used to characterize the free solutions of the resulting extension problem. The solutions have the form $D_* = \Sigma^{\omega_1'} H_* \oplus \Sigma^{\omega_2'} H_*$, where ω_1' and ω_2' are elements of $RO(G)$ satisfying the conditions:

(i) $|(\omega_1')^G| = |V^G|$ and $|\omega_1'| = |\omega_1|$
(ii) $|(\omega_2')^G| = |\omega_1^G|$ and $|\omega_2'| = |V|$, and
(iii) $d_{(V+\omega_1-\omega_1'-\omega_2')} \equiv \pm 1 \mod p$.

The first two of these conditions merely require that ω_1' and ω_2' form a V-shifted ramp associated to the length one ramp ω_1. Let us denote the solution to our extension problem associated to a pair ω_1', ω_2' of elements of $RO(G)$ satisfying these conditions by $D_*[\omega_1', \omega_2']$. Now assume that ω_1'' and ω_2'' are a second pair of elements satisfying these conditions and so providing the solution $D_*[\omega_1'', \omega_2'']$ to our extension problem. Note that our conditions imply that $|(\omega_i')^G| = |(\omega_i'')^G|$ and $|\omega_i'| = |\omega_i''|$ so that $\omega_i' - \omega_i'' \in RO_0(G)$ for $i = 1, 2$. By Theorem 4.4 of **[6]**, $D_*[\omega_1', \omega_2']$ and $D_*[\omega_1'', \omega_2'']$ are isomorphic H_*-modules if and only if

$$d_{\omega_1'-\omega_1''} \equiv d_{\omega_2'-\omega_2''} \equiv \pm 1 \mod p.$$

It follows that, for $p \geq 5$, our algebraic extension problem involving $\widetilde{H}_*^G(B; A)$ has non-isomorphic free solutions. The argument used to prove our freeness results offers no insight into which of these non-isomorphic algebraic solutions is the reduced homology of the space obtained from B by attaching the cell DV. In fact, our argument does not even eliminate the possibility that there could be two attaching maps for the cell DV which produce the same boundary map in the cell-attaching long exact sequence but yield spaces with non-isomorphic homology.

(b) Another way to attempt to gain some insight into our freeness results is to look at the homology of a G-space X localized both at and away from the prime p. For any finite group G, if X is a $\text{Rep}^*(G)$-cell complex containing only even-dimensional cells, then the $RO(G)$-graded equivariant ordinary homology of X with coefficients $A[1/|G|] = A \otimes \mathbb{Z}[1/|G|]$ is a free module over the homology of a point. This is most easily seen by recalling that the ring $A[1/|G|](G/G)$ contains a family of orthogonal idempotents indexed on the conjugacy classes of subgroups of G. These idempotents can be used to split the Mackey functor $A[1/|G|]$ into a product of corresponding factors. Denote the factor associated to the conjugacy class of a subgroup K of G by $A_{(K)}^{1/p}$. It follows from Proposition 3.11 of **[21]** that the $RO(G)$-graded equivariant ordinary homology of a point with $A_{(K)}^{1/p}$ coefficients vanishes in dimension ω unless $|\omega^K| = 0$. Thus, if X is a $\text{Rep}^*(G)$-cell complex containing only even-dimensional cells, then the boundary maps in the cell-attaching long exact sequences used to compute the homology of X with $A_{(K)}^{1/p}$ coefficients must all vanish. Combining these results for the various subsgroups of G, one obtains that the boundary maps in the cell-attaching long exact sequences used to compute the homology of X with $A[1/|G|]$ coefficients also vanish. The homology $H_*^G(X; A[1/|G|])$ of X with $A[1/|G|]$ coefficients is therefore free. It has a set of generators whose dimensions are the dimensions of the cells of X. However, we have a great deal of freedom in picking the dimensions of a set of generators for this free module. In particular, if $G = \mathbb{Z}/p$, then there is one set of generators whose dimensions are those of the cells of X and another set whose dimensions are those of the generators of $H_*^G(B; A)$, whatever they might be. Thus, homology with A$[1/|G|\ ]$ coefficients provides no new information about the dimensions of the generators of $H_*^G(X; A)$.

If the prime p divides the order of the finite group G, then computing the homology of a G-space X with the localization $A_{(p)}$ of A at p as coefficients does not seem to be substantially easier than computing $H_*^G(X;A)$. Thus, for p-local homology, we consider only the case in which $G = \mathbb{Z}/p$. In this case, either by invoking the Universal Coefficient Theorem or by applying the arguments used to prove our freeness results, we obtain that $H_*^G(B;A_{(p)})$ is free over the p-local homology $H_*^{(p)}$ of a point if X is a $\mathrm{Rep}^*(G)$-cell complex satisfying our usual conditions. It is easy to argue that the boundary maps in the cell-attaching long exact sequences used to prove our freeness result are nonzero for $A_{(p)}$ coefficients if and only if they are nonzero for A coefficients. Thus, the dimension shifting that must occur with A coefficients must also occur with $A_{(p)}$ coefficients. There is, however, one slight simplification in the p-local case. For any two elements ω and ω' of $RO(G)$, if $|\omega^G| = |(\omega')^G|$ and $|\omega| = |\omega'|$, then $H_\omega^{(p)} \cong H_{\omega'}^{(p)}$. It follows that the dimension ω of a G/G-generator of a free $H_*^{(p)}$-module is determined only up to the integers $|\omega^G|$ and $|\omega|$. The exact dimensions of the generators of the homology $H_*^G(X;A)$ of a G-space X satisfying the hypotheses of our freeness theorems are therefore not visible in homology localized either at or away from p. The root cause of this is that the Mackey functors $A[d]$, for d prime to p, become isomorphic when localized at or away from p.

CHAPTER 3

An outline of the proof of the main freeness result (Theorem 2.6)

Throughout this chapter, we assume that $G = \mathbb{Z}/p$ and that X is a $\mathrm{Rep}^*(G)$-cell complex formed from even-dimensional cells. We work with the "one cell at a time" filtration $\{X_\alpha\}_{\alpha \in \mathcal{J}}$ of X in which $X_{\alpha+1}$ is formed from X_α by adding a single cell, which is either of the form DV_α for some even-dimensional G-representation V_α or of the form $G \times D^{m_\alpha}$ for some even integer m_α. The compactness axiom for equivariant ordinary homology provides the following result.

LEMMA 3.1. *For any Mackey functor S, the canonical map*

$$\operatorname*{colim}_{\alpha \in \mathcal{J}} H_*^G(X_\alpha; S) \longrightarrow H_*^G(X; S)$$

is an isomorphism.

The proof of Theorem 2.6 therefore consists of two parts. In the first part, we show that, if the homology $H_*^G(B; A)$ of a G-space B is free over H_* with even-dimensional generators and the G-space Y is formed from B by adding a single even-dimensional cell of the form DV or $G \times D^m$, then $H_*^G(Y; A)$ is also free over H_* with even-dimensional generators. In the second part, we assume that the homology $H_*^G(X_\alpha; A)$ of each of the X_α is free over H_*, and argue that $H_*^G(X; A) \cong \operatorname{colim}_{\alpha \in \mathcal{J}} H_*^G(X_\alpha; A)$ is also free over H_*. Since the indexing set $\mathcal{J}$ is an ordinal which may be larger than the ordinal of natural numbers, this second step is actually an inductive argument in which we show that, if β is a limit point of $\mathcal{J}$, then $H_*^G(X_\beta; A) \cong \operatorname{colim}_{\alpha < \beta} H_*^G(X_\alpha; A)$ is free under the assumption that each of the $H_*^G(X_\alpha; A)$ is free.

The second part of this argument is complicated by the difficulty illustrated in Example 2.4. This example shows that, without some additional assumptions, $\operatorname{colim}_{\alpha < \beta} H_*^G(X_\alpha; A)$ need not be free over H_* even if all of the $H_*^G(X_\alpha; A)$ are free. Thus, the heart of the second part of the argument is showing that the finiteness assumption in Theorem 2.6 implies the freeness of this colimit. For this argument, it is necessary to have some fairly detailed information about the behavior of the map $H_*^G(X_\alpha; A) \longrightarrow H_*^G(X_{\alpha+1}; A)$. The need for this extra information adds a certain amount of technical complexity to our first argument dealing with attaching a single cell. The first and second parts of our argument are presented in Sections 3.1 and 3.2, respectively. In Section 3.3, the results from the first two sections are combined to complete the proof of Theorem 2.6. The first part of our argument suffices to prove our freeness result for finite complexes (Theorem 2.1). Thus, anyone whose primary interest is that result may wish to skip from Section 3.1 to Chapter 4, which contains the details of the first part of the argument, before returning to look at Sections 3.2 and 3.3.

3.1. The freeness results for adding a single cell

Throughout this section, we assume that B is a G-space whose homology $H_*^G(B;A)$ is free over H_* with even-dimensional space-like generators. We also assume that the G-space Y is formed from B by adding one even-dimensional cell of the form DV or $G \times D^m$. Associated to this attachment we have a homology cell-attaching long exact sequence of the form

$$\cdots \longrightarrow H_*^G(B;A) \xrightarrow{\chi} H_*^G(Y;A) \xrightarrow{\psi} \widetilde{H}_*^G(S^V;A) \xrightarrow{\partial} H_{*-1}^G(B;A) \longrightarrow \cdots$$

or the form

$$\cdots \longrightarrow H_*^G(B;A) \xrightarrow{\chi} H_*^G(Y;A) \xrightarrow{\psi} \widetilde{H}_*^G(G_+ \wedge S^m;A) \xrightarrow{\partial} H_{*-1}^G(B;A) \longrightarrow \cdots .$$

Our goal is to show that $H_*^G(Y;A)$ is free over H_*. However, because we also need to understand the natural map $\chi : H_*^G(B;A) \longrightarrow H_*^G(Y;A)$, we break the presentation of our freeness argument into two cases. The first of these is trivial. It is presented separately so that its simplicity is not obscured by the notational complexity needed to handle the second case. The proofs of our results for both of these cases are given in Chapter 4.

PROPOSITION 3.2. *Let $G = \mathbb{Z}/p$, and let B be a G-space whose homology $H_*^G(B;A)$ is free over H_* with even-dimensional space-like generators. Assume that the G-space Y is formed from B by adding a single even-dimensional cell and that one of the following conditions holds:*

(i) *the new cell has the form $G \times D^m$, or*
(ii) *the new cell has the form DV and the boundary map ∂ in the associated cell-attaching long exact sequence is zero.*

Then $H_*^G(Y;A)$ is free over H_*. Each of the generators of $H_*^G(B;A)$ is also a generator of $H_*^G(Y;A)$, and $H_*^G(Y;A)$ has one additional generator in the dimension, m or V, of the new cell. Further, the natural map $\chi : H_*^G(B;A) \longrightarrow H_*^G(Y;A)$ is the obvious inclusion.

The central result (Theorem 2.6) in [**12**] gives conditions on V and the dimensions of the generators of $H_*^G(B;A)$ which ensure the vanishing of the boundary map ∂ associated to attaching a cell of the form DV. The critical new insight in [**6**] is that, even if the boundary map ∂ is nonzero, $H_*^G(Y;A)$ still has to be free over H_*. However, in this context, some of the generators of $H_*^G(B;A)$ are shifted in dimension in the process of forming Y from B. The statement of our freeness result for this case is best understood by looking back to Example 2.2 and Definition 2.3. Most of the generators of $H_*^G(B;A)$ pass over to generators of $H_*^G(Y;A)$ in exactly the same dimension. However, finitely many of the G/G-generators of $H_*^G(B;A)$ undergo a dimension shift. The dimensions of the generators in this finite collection form a ramp of the sort introduced in Definition 2.3.

THEOREM 3.3 (Dimension-shifting Theorem). *Let $G = \mathbb{Z}/p$, and let B be a G-space whose homology $H_*^G(B;A)$ is free over H_* with even-dimensional space-like generators. Assume that the G-space Y is formed from B by adding a single even-dimensional cell of the form DV and that the boundary map ∂ in the associated cell-attaching long exact sequence is nonzero. Then $H_*^G(Y;A)$ is free over H_* with even-dimensional space-like generators. All of the G/e-generators and all but a finite set $\mathcal{F}_\partial$ of the G/G-generators of $H_*^G(B;A)$ pass over to generators of $H_*^G(Y;A)$ of the* *same type and in the same dimension. The dimensions of the generators in $\mathcal{F}_\partial$ form a ramp $\omega_1, \omega_2, \ldots, \omega_n$ of length n for some integer $n \geq 1$. Those n generators of $H_*^G(B; A)$ and the new cell DV of Y together produce $n+1$ G/G-generators of $H_*^G(Y; A)$ whose dimensions form a V-shifted ramp $\omega_1', \omega_2', \ldots, \omega_{n+1}'$. The map $\chi : H_*^G(B; A) \longrightarrow H_*^G(Y; A)$ derived from the inclusion $B \subset Y$ takes each of the generators of $H_*^G(B; A)$ not in $\mathcal{F}_\partial$ identically to the corresponding generator of $H_*^G(Y; A)$. Moreover, for $1 \leq i \leq n$, the composite*

$$\Sigma^{\omega_i} H_* \subset H_*^G(B; A) \xrightarrow{\ \chi\ } H_*^G(Y; A) \xrightarrow{\ \pi_i\ } \Sigma^{\omega_i'} H_* \oplus \Sigma^{\omega_{i+1}'} H_*$$

is constructed from one horizontal and one vertical standard shift map. Here, the first map is the inclusion of the summand of $H_*^G(B; A)$ spanned by the generator in $\mathcal{F}_\partial$ of dimension ω_i and the last map is the projection onto the summand of $H_*^G(Y; A)$ spanned by the shifted generators in dimensions ω_i' and ω_{i+1}'.

3.2. Colimits of diagrams of free H_*-modules

As Example 2.4 illustrates, there are fairly nice diagrams of free modules over H_* whose colimit is not free. In this section, we develop the algebraic machinery needed to complete the proof of Theorem 2.6 by introducing a special type of diagram of free H_*-modules whose colimit is free. In the next section, we complete the proof of the theorem by showing that, if X is a G-cell complex satisfying the hypotheses of Theorem 2.6, then the diagram in homology associated to its "one cell at a time" filtration is a diagram of free H_*-modules of this special type.

We are interested in diagrams of free H_*-modules indexed on an ordinal $\mathcal{J}$. Such a diagram consists of a collection $\{C_\alpha\}_{\alpha \in \mathcal{J}}$ of free H_*-modules together with maps $\lambda_{\alpha,\beta} : C_\alpha \longrightarrow C_\beta$, for $\alpha < \beta$ in $\mathcal{J}$, such that the diagram

$$\begin{array}{ccc} C_\alpha & \xrightarrow{\lambda_{\alpha,\beta}} & C_\beta \\ & {}_{\lambda_{\alpha,\gamma}}\searrow & \downarrow{\scriptstyle \lambda_{\beta,\gamma}} \\ & & C_\gamma \end{array}$$

commutes for all $\alpha < \beta < \gamma$ in $\mathcal{J}$. Intuitively, we want to look at ordinal-indexed diagrams of free H_*-modules in which new generators are allowed to appear at any stage of the diagram. However, once a generator appears, it is required to persist, possibly with some dimension shifts, throughout the remainder of the diagram. It is therefore natural to think of generators as belonging to the whole diagram (or, at least, to a terminal part of the diagram) rather than to an individual module appearing in the diagram. For this reason, we want a single global indexing set for our generators with the property that the generators of each module in the diagram are indexed on an appropriate subset of that global set. By adopting the "one cell at a time" approach at the end of Section 1.5, we can use the indexing set $\mathcal{J}$ for this global indexing set.

More precisely, the diagrams of interest to us are those in which, for each $\beta \in \mathcal{J}$, the generators of C_β are indexed on the set $\mathcal{J}(\beta) = \{\alpha \in \mathcal{J} : \alpha < \beta\}$. Thus, we can think of the transition from C_α to $C_{\alpha+1}$ as adding to our diagram one new generator, which can be either type. That generator is to persist throughout the remainder of the diagram as the generator indexed on α and must retain its initial type (G/G or G/e). Denote the dimension of the α-indexed generator of C_β by

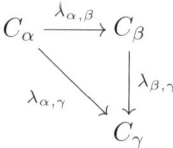

$\omega_{\alpha,\beta}$. This dimension is an element of $RO(G)$ if the generator is of type G/G and an integer if the generator is of type G/e. Ideally, if $\alpha < \beta < \gamma$ in $\mathcal{J}$, then $\omega_{\alpha,\beta}$ and $\omega_{\alpha,\gamma}$ would be equal. However, this restriction would eliminate the possibility of the dimension-shifting that we know occurs in the diagrams of interest to us. Thus, we replace this ideal requirement by a weaker and more complicated set of requirements encoding the sort of dimension-shifting which actually occurs in the context of Theorem 2.6.

DEFINITION 3.4. (a) An ordinal-indexed diagram of free H_*-modules of the sort described above is said to have a consistent set of generators if it satisfies the following conditions for each $\alpha \in \mathcal{J}$:

(i) If the α-indexed generators in the diagram are of type G/e, then $\omega_{\alpha,\beta} = \omega_{\alpha,\gamma}$ for all $\beta, \gamma \in \mathcal{J}$ such that $\alpha < \beta < \gamma$.

(ii) If the α-indexed generators in the diagram are of type G/G, then $|\omega_{\alpha,\beta}| = |\omega_{\alpha,\gamma}|$ and $|\omega^G_{\alpha,\beta}| \geq |\omega^G_{\alpha,\gamma}|$ for all $\beta, \gamma \in \mathcal{J}$ such that $\alpha < \beta < \gamma$. Moreover, if $|\omega^G_{\alpha,\beta}| = |\omega^G_{\alpha,\gamma}|$, then $\omega_{\alpha,\beta} = \omega_{\alpha,\gamma}$.

(iii) If γ is a limit ordinal of $\mathcal{J}$ such that $\alpha < \gamma$ in $\mathcal{J}$, then there is a $\beta \in \mathcal{J}$ such that $\alpha < \beta < \gamma$ and $\omega_{\alpha,\beta} = \omega_{\alpha,\gamma}$

(iv) If $\alpha < \beta < \gamma$ in $\mathcal{J}$ and $\omega_{\alpha,\beta} = \omega_{\alpha,\gamma}$, then $\lambda_{\beta,\gamma}$ takes the α-indexed generator of C_β to the α-indexed generator of C_γ; that is, if $\iota^\alpha_\beta : \Sigma^{\omega_{\alpha,\beta}} H_* \longrightarrow C_\beta$ denotes the inclusion of the summand of C_β spanned by the α-indexed generator, then the diagram

$$\begin{array}{ccc} & \Sigma^{\omega_{\alpha,\beta}} H_* & \\ {}^{\iota^\alpha_\beta}\swarrow & & \searrow^{\iota^\alpha_\gamma} \\ C_\beta & \xrightarrow[\lambda_{\beta,\gamma}]{} & C_\gamma \end{array}$$

commutes.

(v) There is a finite subset $\mathcal{J}_\alpha$ of $\mathcal{J}$ such that, if $\alpha < \beta < \gamma$ in $\mathcal{J}$ and $\omega_{\alpha,\beta} \neq \omega_{\alpha,\gamma}$, then there is a map

$$\tilde{\lambda}^\alpha_{\beta,\gamma} : \Sigma^{\omega_{\alpha,\beta}} H_* \longrightarrow \bigoplus_{\substack{\alpha' \in \mathcal{J}_\alpha \\ \alpha' < \gamma}} \Sigma^{\omega_{\alpha',\gamma}} H_*$$

such that the diagram

$$\begin{array}{ccc} \Sigma^{\omega_{\alpha,\beta}} H_* & \xrightarrow{\tilde{\lambda}^\alpha_{\beta,\gamma}} & \oplus_{\alpha'} \Sigma^{\omega_{\alpha',\gamma}} H_* \\ {\scriptstyle \iota^\alpha_\beta}\downarrow & & \downarrow{\scriptstyle \iota} \\ C_\beta & \xrightarrow{\lambda_{\beta,\gamma}} & C_\gamma \end{array}$$

commutes. Here, the map ι is the inclusion of the summand of C_γ spanned by the generators indexed on the specified subset of $\mathcal{J}_\alpha$.

(b) An ordinal-indexed diagram of free H_*-modules with a consistent set of generators is said to be convergent if, for each $\alpha \in \mathcal{J}$ associated to a G/G-generator, there is a $\beta_0 > \alpha$ such that $\omega_{\alpha,\beta_0} = \omega_{\alpha,\gamma}$ for all $\gamma > \beta_0$. Denote this terminal value of the dimensions associated to α by ω_α. This element ω_α of $RO(G)$ should be thought of as the ultimate dimension of the generators in our diagram indexed on α. In the diagrams of interest to us, all of the dimensions $\omega_{\alpha,\beta}$ are space-like. Thus, if α is an element of $\mathcal{J}$ associated to a G/G-generator, then for $\gamma > \beta > \alpha$, $|\omega^G_{\alpha,\beta}| \geq |\omega^G_{\alpha,\gamma}| \geq 0$. Since the integers $|\omega^G_{\alpha,\beta}|$ decrease with increasing β and are bounded below by 0, the $\beta_0 \in \mathcal{J}$ needed for convergence must exist. If $\alpha \in \mathcal{J}$ is associated to a G/e-generator, then $\omega_{\alpha,\beta} = \omega_{\alpha,\gamma}$ for all $\gamma > \beta > \alpha$, and we take ω_α to be $\omega_{\alpha,\beta}$ for any $\beta > \alpha$.

REMARK 3.5. The fifth condition in Definition 3.4(a) may seem a bit strange. To understand it better, assume that $\alpha < \beta < \gamma$ in $\mathcal{J}$, that α is associated to a G/G-generator, and that $\omega_{\alpha,\beta} \neq \omega_{\alpha,\gamma}$. The image of the α-indexed generator of C_β under the map $\lambda_{\beta,\gamma} : C_\beta \longrightarrow C_\gamma$ could be a multiple of the α-indexed generator of C_γ by an element of H_* in the appropriate dimension. However, it is more likely to be a linear combination of several generators of C_γ. The generators appearing in this linear combination are likely to depend on γ. Condition (v) provides a finite uniform bound on the collections of the generators which appear in these linear combinations.

The misbehavior of the colimit in Example 2.4 arises precisely because no such bound is available. To see this, note that the positive dimensional generators appearing in that example are indexed on the positive integers. The mth generator appears first in $H_*^G(X^m; A)$, where it has a dimension plotting to the point $(2, 2m)$. However, at the very next stage of the filtration, it shifts to a dimension plotting to $(0, 2m)$ and remains at that dimension throughout the remainder of the diagram. For $n > m$, this generator of $H_*^G(X^m; A)$ maps to a nontrivial linear combination of the generators of $H_*^G(X^n; A)$ plotting to the points $(0, 2m)$ and $(2, 2n)$. Thus, the only candidate for a set $\mathcal{J}_m$ satisfying condition (v) in this example would be the infinite set of positive integers greater than or equal to m. The failure of this set to be finite is the source of the failure of the colimit to be free.

For the proof of our algebraic freeness theorem, the subsets $\mathcal{J}_\alpha$ of $\mathcal{J}$ introduced in condition (v) of Definition 3.4(a) need not satisfy any conditions beyond those given there. However, in the application of our algebraic result to the proof of Theorem 2.6, it is essential that these sets have an additional property.

DEFINITION 3.6. The sets $\mathcal{J}_\alpha$ of an ordinal-indexed diagram of free H_*-modules with a consistent set of generators are said to be well positioned if, for each $\alpha' \in \mathcal{J}_\alpha$ and each $\beta \in \mathcal{J}$ such that $\beta > \alpha'$, $|\omega_{\alpha',\beta}| \geq |\omega_{\alpha,\alpha+1}|$ and $|\omega^G_{\alpha',\beta}| \leq |\omega^G_{\alpha,\alpha+1}|$

Our freeness result for diagrams of free H_*-modules is precisely what one would expect based on Definition 3.4.

PROPOSITION 3.7. *Let $\mathcal{J}$ be an ordinal and $\lambda_{\alpha,\beta} : C_\alpha \longrightarrow C_\beta$ be a $\mathcal{J}$-indexed diagram of free H_*-modules having a consistent set of generators which is convergent. Then*

$$C = \operatorname*{colim}_{\beta \in \mathcal{J}} C_\beta$$

is a free H_*-module whose generators are indexed on $\mathcal{J}$. The generator of C indexed on $\alpha \in \mathcal{J}$ is of the same type (G/G or G/e) as the α-indexed generators appearing in the diagram and its dimension is the ultimate dimension ω_α of the α-indexed generators of the diagram. Moreover, if $\alpha < \beta$ and $\omega_{\alpha,\beta} = \omega_\alpha$, then the diagram

$$\begin{array}{ccc} & \Sigma^{\omega_{\alpha,\beta}} H_* = \Sigma^{\omega_\alpha} H_* & \\ {\scriptstyle \iota^\alpha_\beta} \swarrow & & \searrow {\scriptstyle \iota^\alpha} \\ C_\beta & \xrightarrow{\quad \lambda_\beta \quad} & C \end{array}$$

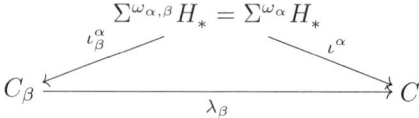

commutes. Here, $\lambda_\beta : C_\beta \longrightarrow C$ and $\iota^\alpha : \Sigma^{\omega_\alpha} H_* \longrightarrow C$ are the canonical map into the colimit and the inclusion of the summand of C spanned by its α-indexed generator, respectively.

PROOF. If $\mathcal{J}$ is not a limit ordinal, then it has a maximal element α_∞. In this case, $C = C_{\alpha_\infty}$, and there is nothing to prove. Thus, we assume that $\mathcal{J}$ is a limit ordinal. For each $\beta \in \mathcal{J}$, let D_β be the summand of C_β spanned by the generators indexed on those $\alpha < \beta$ such that $\omega_{\alpha,\beta} = \omega_\alpha$. If $\gamma > \beta$, then condition (iv) of Definition 3.4 implies that the restriction of $\lambda_{\beta,\gamma} : C_\beta \longrightarrow C_\gamma$ to D_β factors through the inclusion of D_γ into C_γ. Moreover, the resulting map $D_\beta \longrightarrow D_\gamma$ is just the inclusion of a direct summand, basically because the only generators of C_β allowed to appear in D_β are those which undergo no dimension shifting in the part of the diagram beyond β. It follows trivially that $D = \operatorname{colim}_{\beta\in\mathcal{J}} D_\beta$ is a free H_*-module with generators indexed on $\mathcal{J}$. Moreover, the types and dimensions of these generators are exactly those asserted in the proposition for the generators of C. The inclusions $D_\beta \subset C_\beta$ induce a monomorphism $\phi : D \longrightarrow C$. Thus, to show that C is free and has the appropriate generators, it suffices to show that ϕ is an epimorphism. For each $\alpha < \beta$ in $\mathcal{J}$, consider the composite

$$\Sigma^{\omega_{\alpha,\beta}} H_* \xrightarrow{\iota^\alpha_\beta} C_\beta \xrightarrow{\lambda_\beta} C.$$

Taken together, the images of all these maps generate C. Thus, it suffices to show that, for each $\alpha < \beta$ in $\mathcal{J}$, the composite above factors through ϕ. Select $\gamma \in \mathcal{J}$ with $\beta < \gamma$ such that:

(i) $\omega_{\alpha,\gamma} = \omega_\alpha$, and
(ii) for each $\alpha' \in \mathcal{J}_\alpha$, $\alpha' < \gamma$ and $\omega_{\alpha',\gamma} = \omega_{\alpha'}$.

The finiteness of the set $\mathcal{J}_\alpha$ ensures that such a γ exists. Our restrictions on γ imply that, for each $\alpha' \in \mathcal{J}_\alpha$, the generator of C_γ indexed on α' is also a generator of D_γ. The desired factorization then follows from the commutativity of the diagram

$$\begin{array}{ccccc}
\Sigma^{\omega_{\alpha,\beta}} H_* & \xrightarrow{\tilde{\lambda}^\alpha_{\beta,\gamma}} & \bigoplus_{\alpha'} \Sigma^{\omega_{\alpha',\gamma}} H_* & & \\
 & & \downarrow{\scriptstyle \iota} & & \\
\downarrow{\scriptstyle \iota^\alpha_\beta} & & D_\gamma & \longrightarrow & D \\
 & & \downarrow & & \downarrow{\scriptstyle \phi} \\
C_\beta & \xrightarrow{\lambda_{\beta,\gamma}} & C_\gamma & \xrightarrow{\lambda_\gamma} & C.
\end{array}$$

The commutativity of the diagram in the proposition follows easily from the fact that ι^α_β and ι^α factor through D_β and D, respectively. □

3.3. Completing the proof of the main freeness theorem

Here, Proposition 3.7 is employed to complete the proof of Theorem 2.6. Assume that X is a $\mathrm{Rep}^*(G)$-cell complex satisfying the hypotheses of the theorem. Recall the two filtrations on X discussed in Section 1.5. In the first filtration $\{X^n\}_{n\geq 0}$, X^{n+1} is formed from X^n by attaching the collection of cells $\mathcal{J}^{n+1}$. This filtration is used in the definition of a $\mathrm{Rep}^*(G)$-cell complex and in the statement of Theorem 2.6. The second filtration $\{X_\alpha\}_{\alpha\in\mathcal{J}}$ is the "one cell at a time" filtration indexed on an ordinal $\mathcal{J}$. This second filtration is assumed to satisfy the condition that, if $\alpha, \beta \in \mathcal{J}$ and the cells indexed on α and β are in the sets $\mathcal{J}^m$ and $\mathcal{J}^n$, respectively, then $\alpha < \beta$ whenever $m < n$. For technical reasons, an additional condition must be imposed on this filtration. Assume that α and β are elements of $\mathcal{J}$ whose associated cells are of the form DV_α and DV_β and that these two cells lie in the same set $\mathcal{J}^n$. We require that $\alpha < \beta$ if $|V_\alpha^G| < |V_\beta^G|$. Since $|V_\gamma^G| \geq 0$ for all $\gamma \in \mathcal{J}$, it is easy enough to arrange the order of the attachment of cells in the "one cell at a time" filtration so that this extra condition is met.

As in the proof of Proposition 3.7, we may as well assume that $\mathcal{J}$ is a limit ordinal. There are two main obstacles to completing the proof of the theorem by applying the proposition to the $\mathcal{J}$-indexed diagram $\{H_*^G(X_\alpha; A)\}$ of H_*-modules. The first is that, if δ is a limit ordinal in $\mathcal{J}$, then we do not know that $H_*^G(X_\delta; A)$ is a free H_*-module. This is resolved by using the proposition, together with the freeness results from Section 3.1, in a transfinite induction argument on δ. The second problem is that of establishing the existence of the finite sets $\mathcal{J}_\alpha$ required by condition (v) in Definition 3.4(a). We construct these sets as a part of our induction argument.

Throughout our inductive argument on δ, we are going to work with a number of objects which must be indexed on the elements of $\mathcal{J}$. In introducing these indexed objects, we often use an index like $\delta + 1$ when the index δ might seem more natural. The source of this notational clumsiness is that $\mathcal{J}$ typically contains both successor ordinals and limit ordinals. Our inductive argument has to deal with both of these types of ordinals. There is one perfectly reasonable indexing scheme which works very well for the discussion of the successor ordinals, and another completely incompatible scheme which seems quite natural for the discussion of the limit ordinals. Thus, we have selected a third indexing scheme which is uniformly a bit clumsy rather than one of the two that works well for half of the argument but disastrously for the other half.

Our inductive assumption on $\delta \in \mathcal{J}$ is that, for every $\kappa \leq \delta + 1$, the portion of the diagram $\{H_*^G(X_\alpha; A)\}$ of H_*-modules indexed on $\mathcal{J}(\kappa) = \{\alpha \in \mathcal{J} : \alpha < \kappa\}$ is a diagram of free H_*-modules with a consistent set of even-dimensional space-like generators. In the context of this assumption, we denote the finite subsets of $\mathcal{J}$ arising in condition (v) of Definition 3.4(a) by $\mathcal{J}_\alpha(\kappa)$ rather than $\mathcal{J}_\alpha$. We assume that these finite sets are well positioned in the sense of Definition 3.6 and that, for $\kappa < \kappa' \leq \delta+1$, $\mathcal{J}_\alpha(\kappa) \subset \mathcal{J}_\alpha(\kappa')$. Observe that, if the portion of the diagram indexed on $\mathcal{J}(\kappa)$ satisfies (v) for α with respect to the finite set $\mathcal{J}_\alpha(\kappa)$, then it satisfies that condition with respect to any finite set containing $\mathcal{J}_\alpha(\kappa)$. This is important because ultimately we define $\mathcal{J}_\alpha$ to be $\cup_\kappa \mathcal{J}_\alpha(\kappa)$.

One further technical assumption is needed about the dimension $\omega_{\alpha,\alpha+1}$ of a G/G-generator at the point in our diagram where it first appears. Intuitively, this assumption says either that no dimension shifting occurs when the cell DV_α is attached or that the dimension shifting which does occur is tied to at least one cell in a lower filtration $\mathcal{J}^k$ than the filtration $\mathcal{J}^m$ of DV_α. Our precise assumption is that, for each $\alpha \in \mathcal{J}$ associated to a G/G-generator, there is an element $b(\alpha)$ of $\mathcal{J}$, also associated to a G/G-generator, satisfying the conditions:

(i) $|\omega_{\alpha,\alpha+1}| \geq |V_{b(\alpha)}|$ and $|\omega_{\alpha,\alpha+1}^G| \leq |V_{b(\alpha)}^G|$,

(ii) either $b(\alpha) = \alpha$ or the cells $DV_{b(\alpha)}$ and DV_α are in filtrations $\mathcal{J}^k$ and $\mathcal{J}^m$, respectively, such that $k < m$.

We refer to this condition as our bounding assumption on $\omega_{\alpha,\alpha+1}$. The element $b(\alpha)$ is α when the attachment of the cell DV_α causes no dimension shifting. In this case, $\omega_{\alpha,\alpha+1} = V_\alpha$, so the two inequalities in condition (i) are trivially satisfied.

Looking back at Example 2.4 may provide some intuition for this bounding assumption. In that example, the problem with the colimit is not our bounding assumption, which is satisfied, but rather a problem with the finiteness requirement in condition (v) of Definition 3.4(a). Recall that the positive dimensional generators appearing in that example are indexed on the positive integers. For each $m \geq 1$, the mth generator first appears in a dimension plotting to the point $(2, 2m)$. For $m > 1$, this location for the first appearance of the mth generator is the result of a sequence of dimension shifts that begins with a shift involving the first generator at $(2, 2)$. Thus, that bottom cell is the ultimate source of all our dimension shifting, and $b(m) = 1$ for every $m \geq 1$.

We begin our inductive argument by looking at the transition from an element δ of $\mathcal{J}$ to its successor $\delta + 1$. As a part of our inductive assumption, we know that $H_*^G(X_\delta; A)$ is a free H_*-module. If the cell added to construct $X_{\delta+1}$ is of the form $G \times D^{m_\delta}$, for some integer m_δ, or the boundary map ∂ in the cell-attaching long exact sequence associated to the formation of $X_{\delta+1}$ is zero, then Proposition 3.2 applies, and indicates that $H_*^G(X_{\delta+1}; A)$ is a free H_*-module. This result also allows us to take $\omega_{\alpha,\delta+1}$ to be $\omega_{\alpha,\delta}$ for all $\alpha < \delta$ and $\omega_{\delta,\delta+1}$ to be the dimension (m_δ or V_δ) of the cell used to form $X_{\delta+1}$ from X_δ. For each α which is associated to a G/G-generator and is less than δ, we can take $\mathcal{J}_\alpha(\delta+2)$ to be $\mathcal{J}_\alpha(\delta+1)$. If the new generator in $H_*^G(X_{\delta+1}; A)$ is of type G/G, then we begin the process of constructing the set $\mathcal{J}_\delta$ by defining $\mathcal{J}_\delta(\delta+2)$ to be $\{\delta\}$. Given these definitions, it is easy to see that Proposition 3.2 implies that the $\mathcal{J}(\delta+2)$-indexed diagram of H_*-modules is a diagram of free H_*-modules with a consistent set of even-dimensional space-like generators. Our inductive assumptions imply that the finite subsets $\mathcal{J}_\alpha(\delta+2)$ of $\mathcal{J}(\delta+2)$ are well-positioned. Note that, if the generator associated to δ is of type G/G, then $\omega_{\delta,\delta+1} = V_\delta$ so our bounding assumption for $\omega_{\delta,\delta+1}$ is satisfied. This condition must hold for $\alpha < \delta$ by our inductive assumptions.

The case in which the cell attached to construct $X_{\delta+1}$ is of the form DV_δ and the cell-attaching boundary map is nonzero must still be considered. Here, we invoke Theorem 3.3, which asserts that $H_*^G(X_{\delta+1}; A)$ is a free H_*-module, and specifies the dimensions of its generators. Recall that there is a finite set $\mathcal{F}_\partial$ of G/G-generators of $H_*^G(X_\delta; A)$ which do not pass over to generators of $H_*^G(X_{\delta+1}; A)$ in the same dimension. Assume that the elements of $\mathcal{F}_\partial$ are the generators of $H_*^G(X_\delta; A)$ indexed on the elements $\alpha_1, \alpha_2, \ldots, \alpha_n$ of $\mathcal{J}(\delta+1)$. Recall that the dimensions $\omega_1, \omega_2, \ldots, \omega_n$ of these generators form a ramp bounded by V_δ. Also recall that $H_*^G(X_{\delta+1}; A)$ has $n+1$ new generators in dimensions $\omega'_1, \omega'_2, \ldots, \omega'_{n+1}$ that form an associated V_δ-shifted ramp. If $\alpha < \delta$ is not one of the α_i, take the α-indexed generator of $H_*^G(X_{\delta+1}; A)$ to be the obvious one associated by the theorem to the α-indexed generator of $H_*^G(X_\delta; A)$. It follows that $\omega_{\alpha,\delta+1} = \omega_{\alpha,\delta}$. Subject to minor adjustments noted below, take the generator of $H_*^G(X_{\delta+1}; A)$ indexed on α_i, for $1 \leq i \leq n$, to be the new generator in dimension ω'_i so that $\omega_{\alpha_i,\delta+1} = \omega'_i$. Similarly, provisionally take the δ-indexed generator of $H_*^G(X_{\delta+1}; A)$ to be the new generator in dimension ω'_{n+1}.

Since $\omega_{\delta,\delta+1} = \omega'_{n+1}$, the conditions $|\omega_{\delta,\delta+1}| > |\omega_n|$ and $|\omega^G_{\delta,\delta+1}| = |\omega^G_n|$ are satisfied. But ω_n is $\omega_{\alpha_n,\delta}$, which satisfies the conditions $|\omega_{\alpha_n,\delta}| = |\omega_{\alpha_n,\alpha_n+1}|$ and $|\omega^G_{\alpha_n,\delta}| \leq |\omega^G_{\alpha_n,\alpha_n+1}|$. It follows that $b(\alpha_n)$ is an obvious choice for the bound $b(\delta)$ of δ. The only difficulty which might arise from this choice is that the lower filtration condition in our bounding assumption might fail. If $b(\alpha) \neq \alpha$, it obviously doesn't fail. In the case $b(\alpha) = \alpha$, we can use the technical assumption on the "one cell at a time" filtration imposed at the beginning of this section to show that, since

$$|V_\delta^G| < |\omega^G_{\delta,\delta+1}| \leq |\omega^G_{\alpha_n,\alpha_n+1}| = |V^G_{\alpha_n}|,$$

DV_δ is in a higher filtration $\mathcal{J}^m$ than DV_{α_n}.

It should now be easy to see that conditions (i), (ii), and (iii) of Definition 3.4(a) are satisfied for the diagram $\{H^G_*(X_\alpha; A)\}$ of H_*-modules indexed on $\mathcal{J}(\delta+2)$. Moreover, the description of the map $\chi : H^G_*(X_\delta; A) \longrightarrow H^G_*(X_{\delta+1}; A)$ given in Theorem 3.3 implies that condition (iv) is also satisfied. To complete our proof that the collection $\{H^G_*(X_\alpha; A)\}_{\alpha \leq \delta+1}$ is a $\mathcal{J}(\delta+2)$-indexed diagram of free H_*-modules having a consistent set of space-like generators, we must construct the finite sets $\mathcal{J}_\alpha(\delta+2)$ and show that condition (v) of Definition 3.4(a) is satisfied for the part of our homology diagram indexed on $\mathcal{J}(\delta+2)$. The set $\{\delta\}$ is a natural choice for $\mathcal{J}_\delta(\delta+2)$ and is obviously well positioned.

If $\alpha < \delta$ is associated to a G/G-generator and none of the α_i are in $\mathcal{J}_\alpha(\delta+1)$, then we can take $\mathcal{J}_\alpha(\delta+2)$ to be $\mathcal{J}_\alpha(\delta+1)$. By our inductive assumption, this set is well positioned. Moreover, it follows easily from Theorem 3.3 that this set suffices to ensure that condition (v) is satisfied with respect to α for the part of the diagram indexed on the set $\mathcal{J}(\delta+2)$.

To define $\mathcal{J}_\alpha(\delta+2)$ for those α such that $\mathcal{J}_\alpha(\delta+1)$ contains at least one of the α_i, we must examine the restriction of the map $\chi : H^G_*(X_\delta; A) \longrightarrow H^G_*(X_{\delta+1}; A)$ to the summand of $H^G_*(X_\delta; A)$ spanned by each generator indexed on one of the α_i in $\mathcal{J}_\alpha(\delta+1)$. This examination may indicate that we need to adjust the generators of $H^G_*(X_{\delta+1}; A)$ indexed on the α_i and δ. It is important to note that these adjustments do not involve a change in the dimension. These adjustments are best described by adopting the notational convention that δ is α_{n+1}.

Consider the composite

$$\Sigma^{\omega_i} H_* \subset H^G_*(X_\delta; A) \xrightarrow{\chi} H^G_*(X_{\delta+1}; A) \xrightarrow{\pi^i} \Sigma^{\omega'_i} H_* \oplus \Sigma^{\omega'_{i+1}} H_*$$

in which the first map is the inclusion of the summand of $H^G_*(X_\delta; A)$ spanned by the generator indexed on α_i and the last map is the projection onto the summand of $H^G_*(X_{\delta+1}; A)$ spanned by the generators indexed on α_i and α_{i+1}. Theorem 3.3 indicates that this composite is constructed from standard shift maps. Clearly, if $\alpha_i \in \mathcal{J}_\alpha(\delta+1)$, then α_{i+1} must be added to the set $\mathcal{J}_\alpha(\delta+1)$ in the process of forming $\mathcal{J}_\alpha(\delta+2)$. However, it is possible that even more indices must be added. The possible adjustment in the generators of $H^G_*(X_{\delta+1}; A)$ mentioned above provides us with some control over which indices must be added.

The composite

$$\Sigma^{\omega_i} H_* \subset H^G_*(X_\delta; A) \xrightarrow{\chi} H^G_*(X_{\delta+1}; A),$$

which we denote by χ^i, is completely determined by the image of the standard element μ of $A(G/G) = (\Sigma^{\omega_i} H_*)_{\omega_i}(G/G)$. This image $(\chi^i_{\omega_i})(G/G)(\mu)$ must lie in a summand of $H^G_*(X_{\delta+1}; A)$ spanned by a finite number of generators. Pick a minimal set of generators whose span contains this element. Denote the indices of these generators by $\beta_1, \beta_2, \ldots, \beta_m$ and the dimensions of these generators by ω''_1, $\omega''_2, \ldots, \omega''_m$. If we did not need to show that the set $\mathcal{J}_\alpha(\delta+2)$ is well positioned, then we could just add the indices $\beta_1, \beta_2, \ldots, \beta_m$ into $\mathcal{J}_\alpha(\delta+1)$ in the process of forming $\mathcal{J}_\alpha(\delta+2)$ whenever $\alpha_i \in \mathcal{J}_\alpha(\delta+1)$. Doing this for all α and i would produce finite sets $\mathcal{J}_\alpha(\delta+2)$ satisfying condition (v) for the part of our homology diagram indexed on $\mathcal{J}(\delta+2)$. However, in order to ensure that the set $\mathcal{J}_\alpha(\delta+2)$ is well positioned, we must be a bit more careful about what we add to $\mathcal{J}_\alpha(\delta+1)$.

From the description of H_* given in Proposition 1.8, it is easy to see that, for each j, the dimension ω_j'' must satisfy one of the following three conditions:

(i) $|\omega_j''| \geq |\omega_i|$ and $|(\omega_j'')^G| \leq |\omega_i^G|$,
(ii) $|\omega_j''| = |\omega_i|$ and $|(\omega_j'')^G| > |\omega_i^G|$, or
(iii) $|\omega_j''| < |\omega_i|$ and $|(\omega_j'')^G| = |\omega_i^G|$.

Note that, since α_i is assumed to be in the well-positioned set $\mathcal{J}_\alpha(\delta+1)$, ω_i must satisfy the conditions

$$|\omega_i| \geq |\omega_{\alpha,\alpha+1}| \quad \text{and} \quad |\omega_i^G| \leq |\omega_{\alpha,\alpha+1}^G|.$$

Thus, if ω_j'' satisfies the first of the three conditions above, then adding the associated index to $\mathcal{J}_\alpha(\delta+1)$ will not prevent the set $\mathcal{J}_\alpha(\delta+2)$ from being well positioned. However, if ω_j'' satisfies either of the other two conditions, then we cannot afford to add the associated index to $\mathcal{J}_\alpha(\delta+1)$. At this point, it becomes important that the composite $\pi^i \circ \chi^i$ is constructed from standard shift maps. This implies that, if ω_j'' satisfies the second of the conditions above, then by adding an appropriate multiple of the generator of $H_*^G(X_{\delta+1}; A)$ indexed on β_j to the generator indexed on α_i, we can eliminate the need to include the β_j-indexed generator in the list of those required to span the minimal summand containing $(\chi^i_{\omega_i})(G/G)(\mu)$. The desired multiple is, of course, obtained by multiplying by some element of $H_{\omega_i'-\omega_j''}(G/G)$. Similarly, if ω_j'' satisfies the third condition, then the generator of $H_*^G(X_{\delta+1}; A)$ indexed on α_{i+1} can be adjusted by adding a multiple of the β_j-indexed generator to eliminate the need for that β_j-indexed generator in the spanning set for this minimal summand.

The one difficulty which might arise in this process comes from the fact that the generator of $H_*^G(X_{\delta+1}; A)$ indexed on α_{i+1} must be adjusted to control the spanning sets for the images of both the α_i- and α_{i+1}-indexed generators of $H_*^G(X_\delta; A)$. However, the ramp arrangement of the generators of $H_*^G(X_\delta; A)$ indexed on $\mathcal{F}_\partial$ ensures that the adjustments made for each of these two generators of $H_*^G(X_\delta; A)$ are completely invisible to the other generator. Thus, the desired adjustment to the basis for the free H_*-module $H_*^G(X_{\delta+1}; A)$ can be made. This ensures that we need not add the index β_j to $\mathcal{J}_\alpha(\delta+1)$ unless ω_j'' satisfies the first of our three conditions. It follows that there is a finite well positioned set $\mathcal{J}_\alpha(\delta+2)$ satisfying condition (v) for the part of our homology diagram indexed on $\mathcal{J}(\delta+2)$. This completes the part of our inductive argument dealing with the transition from δ to its successor $\delta+1$.

Now we must verify our induction assumptions for a limit ordinal δ of $\mathcal{J}$. The first step in verifying our assumptions for δ is showing that the part of our homology diagram indexed on $\mathcal{J}(\delta)$ has a consistent set of even-dimensional space-like generators. Given this, Proposition 3.7 indicates that $H_*^G(X_\delta; A) \cong \operatorname{colim}_{\alpha<\delta} H_*^G(X_\alpha; A)$ is a free H_*-module and specifies the dimensions of its generators. We must then show that the part of our homology diagram indexed on $\mathcal{J}(\delta+1)$ has a consistent set of even-dimensional space-like generators and satisfies all our other inductive assumptions.

Since δ is a limit ordinal in $\mathcal{J}$, it is easy to see that conditions (i) to (iv) of Definition 3.4(a) are satisfied for the part of our homology diagram indexed on $\mathcal{J}(\delta)$. The finite sets $\mathcal{J}_\alpha(\delta)$ with respect to which that part of our diagram ought to satisfy condition (v) are given by $\mathcal{J}_\alpha(\delta) = \cup_{\gamma<\delta}\mathcal{J}_\alpha(\gamma)$. The obvious difficulty with these sets is that it is not clear that they are finite. It is however, easy to see that, if they are finite, then condition (v) is satisfied with respect to them on the part of our diagram indexed on $\mathcal{J}(\delta)$. Moreover, the set $\mathcal{J}_\alpha(\delta)$ is obviously well positioned since each of the $\mathcal{J}_\alpha(\gamma)$ is. By our inductive assumptions, if $\beta < \gamma < \delta$, then $\mathcal{J}_\alpha(\beta) \subset \mathcal{J}_\alpha(\gamma)$. Moreover, these two sets are usually equal. New elements are added only in a transition from a set of the form $\mathcal{J}_\alpha(\beta+1)$ to the set $\mathcal{J}_\alpha(\beta+2)$ and only in the special case in which one of the generators indexed on $\mathcal{J}_\alpha(\beta+1)$ is involved in the dimension shifting occurring in the transition from $H_*^G(X_\beta; A)$ to $H_*^G(X_{\beta+1}; A)$. Thus, it suffices to show that there are only finitely many dimension shifts which can insert elements into $\mathcal{J}_\alpha(\delta)$. Ultimately, this follows from the finiteness assumption in the hypotheses of Theorem 2.6. Our bounding assumption on $\omega_{\alpha,\alpha+1}$ and the fact that $\mathcal{J}_\alpha(\delta)$ is well positioned are the means by which this finiteness assumption is exploited.

Assume that $\alpha' \in \mathcal{J}_\alpha(\beta+1)$ is involved in a dimension shift when $X_{\beta+1}$ is formed from X_β. The cell added in this transition must be of the form DV_β. Moreover, since the dimension-shifting affects the generator of $H_*^G(X_\beta; A)$ indexed on α', its dimension $\omega_{\alpha',\beta}$ must satisfy the conditions

$$|V_\beta| > |\omega_{\alpha',\beta}| \text{ and } |V_\beta^G| < |\omega_{\alpha',\beta}^G|.$$

Since $\mathcal{J}_\alpha(\beta+1)$ is well positioned and contains α', the dimension $\omega_{\alpha',\beta}$ must also satisfy the conditions

$$|\omega_{\alpha',\beta}| \geq |\omega_{\alpha,\alpha+1}| \text{ and } |\omega_{\alpha',\beta}^G| \leq |\omega_{\alpha,\alpha+1}^G|.$$

Our bounding assumption for $\omega_{\alpha,\alpha+1}$ provides an element $b(\alpha)$ of $\mathcal{J}$ associated to a G/G-generator such that

$$|\omega_{\alpha,\alpha+1}| \geq |V_{b(\alpha)}| \text{ and } |\omega_{\alpha,\alpha+1}^G| \leq |V_{b(\alpha)}^G|.$$

Combining these inequalities, we see that

$$|V_\beta| > |V_{b(\alpha)}| \text{ and } |V_\beta^G| < |V_{b(\alpha)}^G|.$$

We would like to use the finiteness assumption in the hypotheses of Theorem 2.6 to argue that there are only finitely many β for which these last two inequalities hold. This would imply that, in the formation of $\mathcal{J}_\alpha(\delta)$, there are only finitely many times when we can add elements. Since only finitely many elements can be added whenever elements are added, it would follow that $\mathcal{J}_\alpha(\delta)$ is finite.

To invoke the finiteness assumption in Theorem 2.6, we must show that the the filtration $\mathcal{J}^k$ of $DV_{b(\alpha)}$ is lower than the filtration $\mathcal{J}^n$ of DV_β. Note that the filtration $\mathcal{J}^n$ of DV_β is at least as high as that of DV_α since DV_β is added after DV_α. Moreover, by our bounding assumption, either the filtration $\mathcal{J}^k$ of $DV_{b(\alpha)}$ is lower than the filtration $\mathcal{J}^m$ of DV_α or $b(\alpha) = \alpha$. Thus, unless $b(\alpha) = \alpha$, the required filtration condition holds. If $b(\alpha) = \alpha$, then

$$|V_\beta| > |V_\alpha| \text{ and } |V_\beta^G| < |V_\alpha^G|.$$

However, we ordered the cells of X in such a way that this condition cannot hold if DV_β and DV_α are in the same filtration. Thus, even in this case, the filtration $\mathcal{J}^k$ of $DV_{b(\alpha)} = DV_\alpha$ is lower than the filtration of DV_β.

This completes our proof that the portion of our homology diagram indexed on $\mathcal{J}(\delta)$ has a consistent set of even-dimensional space-like generators. From this, we conclude that $H_*^G(X_\delta; A)$ is a free H_*-module with even-dimensional space-like generators. It follows easily from Proposition 3.7 that the portion of our homology diagram indexed on $\mathcal{J}(\delta + 1)$ satisfies conditions (i) through (iv) of Definition 3.4(a). By taking $\mathcal{J}_\alpha(\delta + 1)$ to be $\mathcal{J}_\alpha(\delta)$ for each α associated to a G/G-generator of $H_*^G(X_\delta; A)$, it is easy to see that condition (v) is also satisfied. The set $\mathcal{J}_\alpha(\delta+1)$ is, of course, well positioned since $\mathcal{J}_\alpha(\delta)$ is. Moreover, our bounding assumption is satisfied for each α indexing a G/G-generator of $H_*^G(X_\delta; A)$. Thus, our inductive assumptions are satisfied for the limit ordinal δ.

The last step in the proof of Theorem 2.6 is showing that our entire homology diagram has a consistent set of even-dimensional space-like generators. Conditions (i) through (iv) of Definition 3.4(a) are obviously satisfied since each instance of them only refers to a portion of the diagram indexed on some subset $\mathcal{J}(\delta)$ of $\mathcal{J}$. For condition (v), we take $\mathcal{J}_\alpha$ to be $\cup_\beta \mathcal{J}_\alpha(\beta)$, where β runs over the elements of $\mathcal{J}$ larger than α. As in the part of our inductive argument dealing with a limit ordinal δ of $\mathcal{J}$, if we can show that $\mathcal{J}_\alpha$ is finite, it then follows that condition (v) is satisfied. The argument for the finiteness of $\mathcal{J}_\alpha$ is essentially identical to the one given for a limit ordinal δ, and so is not repeated. Proposition 3.7 can now be invoked to complete the proof of Theorem 2.6.

CHAPTER 4

Proving the single-cell freeness results

Throughout this chapter, B is assumed to be a G-space whose homology is free over H_* with even-dimensional space-like generators. We also assume that the G-space Y is formed from B by adding a single even-dimensional cell of the form DV or $G \times D^m$. Associated to this attachment we have a homology cell-attaching long exact sequence of the form

$$\cdots \longrightarrow H_*^G(B;A) \xrightarrow{\chi} H_*^G(Y;A) \xrightarrow{\psi} \widetilde{H}_*^G(S^V;A) \xrightarrow{\partial} H_{*-1}^G(B;A) \longrightarrow \cdots$$

or the form

$$\cdots \longrightarrow H_*^G(B;A) \xrightarrow{\chi} H_*^G(Y;A) \xrightarrow{\psi} \widetilde{H}_*^G(G_+ \wedge S^m;A) \xrightarrow{\partial} H_{*-1}^G(B;A) \longrightarrow \cdots .$$

Our goal is to prove Proposition 3.2 and Theorem 3.3. The proposition follows trivially from the following two results.

LEMMA 4.1. *Assume that the homology $H_*^G(B;A)$ of B is free over H_* with even-dimensional space-like generators. If the cell attached to B is of the form $G \times D^m$, then the boundary map ∂ in the cell-attaching long exact sequence is zero.*

LEMMA 4.2. *If the boundary map ∂ in the cell-attaching long exact sequence is zero, then either*

$$H_*^G(Y;A) \cong H_*^G(B;A) \oplus \widetilde{H}_*^G(S^V;A)$$

or

$$H_*^G(Y;A) \cong H_*^G(B;A) \oplus \widetilde{H}_*^G(G_+ \wedge S^m;A),$$

depending on which type of cell is added to B in the formation of Y. Moreover, under this isomorphism, the natural map $\chi : H_*^G(B;A) \longrightarrow H_*^G(Y;A)$ is identified with the inclusion of $H_*^G(B;A)$ into the direct sum as the first summand. Thus, if $H_*^G(B;A)$ is free over H_*, then $H_*^G(Y;A)$ is free over H_* with generators consisting of the generators of $H_*^G(B;A)$ and one additional G/G-generator.

The first of these lemmas follows directly from Lemma 1.12(b), which indicates that there can be no nontrivial maps from $\widetilde{H}_*^G(G_+ \wedge S^m;A)$ to $H_{*-1}^G(B;A)$. The second follows from the projectivity of the H_*-modules $\widetilde{H}_*^G(S^V;A) \cong \Sigma^V H_*$ and $\widetilde{H}_*^G(G_+ \wedge S^m;A) \cong \Sigma^m (H_*)_{G/e}$.

The remainder of this chapter, and both of the next two chapters, are devoted to the proof of Theorem 3.3. Because of the length of this proof, the next section provides a quick overview of the argument. Modulo the proofs of some key technical results, the details of that argument are then presented in the remaining three sections of this chapter. The proofs of those technical results are rather lengthy, and are therefore given separately in the next two chapters.

4.1. A proof overview for the dimension-shifting theorem (Theorem 3.3)

For the remainder of this chapter, we assume that the cell attached to the G-space B to form the G-space Y is of the form DV and that the boundary map ∂ in the associated cell-attaching long exact sequence is nonzero. In this context, some dimension shifting must occur in the transition from $H_*^G(B;A)$ to $H_*^G(Y;A)$. The role of the set $\mathcal{F}_\partial$ appearing in Theorem 3.3 is to keep track of that shifting. Let J be the summand of $H_*^G(B;A)$ spanned by the generators in $\mathcal{F}_\partial$ (which are all of type G/G), and Z be the summand spanned by all the other generators (of both type G/G and type G/e) so that $H_*^G(B;A) \cong J \oplus Z$. The set $\mathcal{F}_\partial$ can be chosen to ensure that the composite

$$Z \subset H_*^G(B;A) \xrightarrow{\chi} H_*^G(Y;A)$$

is a monomorphism.

Define the quotient Q of $H_*^G(Y;A)$ by the short exact sequence

$$0 \longrightarrow Z \longrightarrow H_*^G(Y;A) \xrightarrow{\pi} Q \longrightarrow 0.$$

An appropriate choice of the set $\mathcal{F}_\partial$ also allows us to construct a long exact sequence

$$\cdots \longrightarrow J \xrightarrow{\chi'} Q \xrightarrow{\psi'} \widetilde{H}_*^G(S^V;A) \xrightarrow{\partial'} \Sigma J \longrightarrow \cdots$$

for Q from the original cell-attaching long exact sequence for $H_*^G(Y;A)$. This new long exact sequence is essentially identical to the cell-attaching long exact sequence associated to the special case discussed in Example 2.2. This new sequence has the advantage of being considerably simpler than the one from which it is constructed — enough so that we can actually compute Q in some critical dimensions. Observe that, if Q is a free H_*-module, then its defining short exact sequence splits, yielding an isomorphism

$$H_*^G(Y;A) \cong Z \oplus Q.$$

From this, the freeness of $H_*^G(Y;A)$ follows immediately. By looking a bit more carefully at this isomorphism, we can also verify the claims of the theorem about the natural map $\chi : H_*^G(B;A) \longrightarrow H_*^G(Y;A)$.

We have now reduced the proof of Theorem 3.3 to showing that Q is free over H_* on an appropriate set of generators. Recall that the dimensions of the n generators of $H_*^G(B;A)$ in $\mathcal{F}_\partial$ form a ramp $\omega_1, \omega_2, \ldots, \omega_n$ and that the generators of Q ought to be in dimensions $\omega_1', \omega_2', \ldots, \omega_{n+1}'$ forming a V-shifted ramp of the sort introduced in Definition 2.3. It follows easily from the values of H_* given in Proposition 1.8 that $Q_{\omega_i'}$ should be isomorphic to A for $1 \leq i \leq n+1$. The first step in showing that Q is free is using the long exact sequence above to verify that, if the dimensions ω_i' are appropriately chosen, then Q is isomorphic to A in these dimensions. This step is completed in Proposition 4.12. The proof of this result is a nontrivial computation, the details of which are presented in Chapter 5. Computing Q in these dimensions allows us to construct a map $\theta : J' \longrightarrow Q$ comparing Q with a free H_*-module $J' = \bigoplus_{1\leq i\leq n+1} \Sigma^{\omega_i'} H_*$ having the appropriate generators.

To show that θ is an isomorphism, we wish to insert J' into a long exact sequence comparable to our long exact sequence for Q. This can be accomplished by lifting the map $\chi' : J \longrightarrow Q$ through the map $\theta : J' \longrightarrow Q$. Constructing the lifting $\bar{\chi} : J \longrightarrow J'$ which makes the diagram

$$\begin{array}{ccc} & & J' \\ & \overset{\bar{\chi}}{\nearrow} & \downarrow \theta \\ J & \xrightarrow[\chi']{} & Q \end{array}$$

commute requires the computation of the values of Q in some additional dimensions. These computations are described in Proposition 4.14, whose proof is given in Chapter 5.

The lifting $\bar{\chi}$ allows us to construct the commuting diagram

$$\begin{array}{ccccccccc} \cdots \longrightarrow & J & \xrightarrow{\bar{\chi}} & J' & \xrightarrow{\bar{\psi}} & \widetilde{H}_*^G(S^V; A) & \xrightarrow{\bar{\partial}} & \Sigma J & \longrightarrow \cdots \\ & \downarrow = & & \downarrow \theta & & \downarrow = & & \downarrow = & \\ \cdots \longrightarrow & J & \xrightarrow[\chi']{} & Q & \xrightarrow[\psi']{} & \widetilde{H}_*^G(S^V; A) & \xrightarrow[\partial']{} & \Sigma J & \longrightarrow \cdots \end{array}$$

in which $\bar{\psi}$ is defined to be $\psi' \circ \theta$. If we knew that the top row of this diagram were a long exact sequence, it would follow immediately that θ is an isomorphism. Fortunately, it is possible to give fairly simple conditions for the exactness of a sequence of free H_*-modules of this form. These exactness criteria are presented in Section 4.3 and proven in Chapter 6. The precise existence results for the maps θ and $\bar{\chi}$ stated in Section 4.4 include the information needed to show that the top row in the diagram above satisfies these exactness conditions.

The next three sections are devoted to filling in the details of this quick sketch of the proof of Theorem 3.3. The first of these sections discusses the selection of the subset $\mathcal{F}_\partial$ of $\mathcal{F}$ and the construction of our long exact sequence characterizing the quotient Q of $H_*^G(Y; A)$. The second of these is devoted to the presentation of our exactness criteria for sequences of free H_*-modules like the top row of the last diagram above. This is somewhat out of order with regard to the sketch given above. However, being aware of the precise criteria for exactness makes it easier to appreciate the detailed results about the values of Q and behavior of the maps θ and χ presented in the third section. That third section also contains a wrap-up of the proof of Theorem 3.3.

4.2. Simplifying the cell-attaching long exact sequence

In this section, we retain the assumptions about B and Y made in the previous section. As we noted there, the assumption that the cell-attaching boundary map ∂ is nonzero forces some dimension shifting in the transition from $H_*^G(B; A)$ to $H_*^G(Y; A)$. This shifting is extremely hard to understand unless the dimensions of the generators of $H_*^G(B; A)$ hit by the map ∂ form a ramp like that in Figure 2.1. Fortunately, a rather minimal adjustment of our set of generators for $H_*^G(B; A)$ ensures that those dimensions do so. Fundamentally, the function of the set $\mathcal{F}_\partial$ in Theorem 3.3 is to keep track of this change of basis for $H_*^G(B; A)$. Once this change of basis has been made, it is easy to see that the composite

$$Z \subset H_*^G(B; A) \xrightarrow{\chi} H_*^G(Y; A)$$

is a monomorphism and to construct our long exact sequence for the quotient Q of $H_*^G(Y; A)$.

The desired change of basis is best understood by looking at the following example, which illustrates how that change of basis is accomplished in the simplest possible cases.

EXAMPLE 4.3. Assume that the boundary map $\partial : \widetilde{H}_*^G(S^V; A) \longrightarrow H_{*-1}^G(B; A)$ factors through the summand of $H_*^G(B; A)$ spanned by two generators in dimensions ω_1 and ω_2 satisfying

$$|\omega_1| \leq |\omega_2| < |V| \qquad \text{and} \qquad |\omega_1^G| \geq |\omega_2^G| > |V^G|.$$

Since $\widetilde{H}_*^G(S^V; A)$ is a free H_*-module on one G/G-generator in dimension V, the map ∂ is completely determined by its behavior in dimension V. The two generators in dimensions ω_1 and ω_2 each contribute a copy of $\langle \mathbb{Z}/p \rangle$ to $H_{V-1}^G(B; A)$, so the map ∂_V has the form

$$A \xrightarrow{\partial_V} \langle \mathbb{Z}/p \rangle \oplus \langle \mathbb{Z}/p \rangle \subset H_{V-1}^G(B; A),$$

and is completely determined by the image of the standard generator $\mu \in A(G/G)$ of A. Let $\partial_V(\mu) = (x, y) \in (\langle \mathbb{Z}/p \rangle \oplus \langle \mathbb{Z}/p \rangle)(G/G)$, and assume that both x and y are nonzero. We want to define $\Lambda : H_*^G(B; A) \longrightarrow H_*^G(B; A)$ so that $\Lambda_{V-1} \circ \partial_V(\mu) = (x, 0)$; that is, so that Λ pulls the boundary map off of the generator in dimension ω_2. Clearly, we want to define Λ to be the identity on all the generators other than our two special ones, and to define it on those two generators so that, in $H_{V-1}^G(B; A)$, it takes $(a, b) \in (\langle \mathbb{Z}/p \rangle \oplus \langle \mathbb{Z}/p \rangle)(G/G)$ to $(a, b - x^{-1}ya) \in (\langle \mathbb{Z}/p \rangle \oplus \langle \mathbb{Z}/p \rangle)(G/G)$. This formula suggests that Λ should also be the identity on the generator in dimension ω_2, and should take the generator in dimension ω_1 to some linear combination of itself and the generator in dimension ω_2.

The generator in dimension ω_1 contributes a copy of A to $H_{\omega_1}^G(B; A)$. Denote the standard generating element of this copy of A by μ_1. The generator in dimension ω_2 contributes one of the Mackey functors $A[d]$, $\langle \mathbb{Z} \rangle$, R, or $\langle \mathbb{Z}/p \rangle$ to $H_{\omega_1}^G(B; A)$, depending on the relative positions of ω_1 and ω_2 in the usual plot of elements of $RO(G)$. This contribution contains a generating element of one of the forms μ_2, ϵ, ξ, or $\epsilon\xi$. Define Λ on the generator in dimension ω_1 by

$$\Lambda(\mu_1) = \begin{cases} \mu_1 + c\mu_2 & \text{if } |\omega_1| = |\omega_2| \text{ and } |\omega_1^G| = |\omega_2^G|, \\ \mu_1 + c\epsilon & \text{if } |\omega_1| < |\omega_2| \text{ and } |\omega_1^G| = |\omega_2^G|, \\ \mu_1 + c\xi & \text{if } |\omega_1| = |\omega_2| \text{ and } |\omega_1^G| > |\omega_2^G|, \\ \mu_1 + c\epsilon\xi & \text{if } |\omega_1| < |\omega_2| \text{ and } |\omega_1^G| > |\omega_2^G|. \end{cases}$$

Here, c is an integer which can be selected to ensure that Λ behaves as desired in dimension $V - 1$. Obviously, Λ is an isomorphism of H_*-modules.

This example shows that we can push ∂ off of one generator onto any other generator which plots to the same point or to a point below and/or to the right. Essentially, by pushing ∂ off of as many generators as possible, we can push it onto a finite set $\mathcal{F}_\partial$ of G/G-generators of $H_*^G(B; A)$ whose dimensions form a ramp. Our precise definition of $\mathcal{F}_\partial$ is easily understood when viewed in terms of this pushing off process.

DEFINITION 4.4. Let B be a G-space whose homology $H_*^G(B; A)$ is free over H_* with even-dimensional space-like generators. Assume that the G-space Y is formed from B by adding a single even-dimensional cell of the form DV, and that the boundary map

$$\partial : \widetilde{H}^G_*(S^V; A) \longrightarrow H^G_{*-1}(B; A)$$

in the associated cell-attaching long exact sequence is nonzero. This map is completely determined by ∂_V, which has the form

$$\partial_V : A \longrightarrow \oplus\langle \mathbb{Z}/p \rangle,$$

where the direct sum is indexed on those G/G-generators of $H^G_*(B; A)$ lying in a dimension ω bounded by V — that is, satisfying

$$|\omega| < |V| \qquad \text{and} \qquad |\omega^G| > |V^G|.$$

Moreover, if $\mu \in A(G/G)$ is the standard generator of A, then ∂_V is completely determined by $\partial_V(\mu)$, which has only finitely many nonzero coordinates in the direct sum $(\oplus\langle \mathbb{Z}/p \rangle)(G/G)$. Let $\mathcal{F}$ be the set of G/G-generators of $H^G_*(B; A)$ and $\mathcal{F}_1$ be the subset of $\mathcal{F}$ consisting of those generators corresponding to the nonzero coordinates of $\partial_V(\mu)$. Since the generators of $H^G_*(B; A)$ are space-like, the dimension ω of any one of them satisfies $0 \leq |\omega^G| \leq |\omega|$. Thus, there is a minimum value for $|\omega|$ among the dimensions ω of the generators in $\mathcal{F}_1$. Among all the generators in $\mathcal{F}_1$ with this minimum value for $|\omega|$, select one for which $|\omega^G|$ is maximal. This generator is the first element of $\mathcal{F}_\partial$; denote its dimension by ω_1. Note that $|\omega_1| < |V|$ and $|V^G| < |\omega_1^G|$ since ∂_V is nonzero on the selected generator.

Now assume that the first i elements of $\mathcal{F}_\partial$ have been selected and that their dimensions $\omega_1, \omega_2, \ldots, \omega_i$ form a ramp of length i. Let $\mathcal{F}_{i+1}$ be the subset of $\mathcal{F}_1$ consisting of those generators having a dimension ω satisfying $|\omega^G| > |\omega_i^G|$. Our selection process will ensure that the dimension ω of any generator in $\mathcal{F}_{i+1}$ also satisfies $|\omega| > |\omega_i|$. If the set $\mathcal{F}_{i+1}$ is nonempty, then there is a minimum value for $|\omega|$ among the dimensions ω of the generators in $\mathcal{F}_{i+1}$. Among all the generators in $\mathcal{F}_{i+1}$ with this minimum value for $|\omega|$, select one for which $|\omega^G|$ is maximal. This generator is the $(i+1)^{\text{st}}$ element of $\mathcal{F}_\partial$; denote its dimension by ω_{i+1}. Since the sets $\mathcal{F}_i$ are finite and decreasing in size, this inductive process eventually stops at an integer n for which the set $\mathcal{F}_{n+1}$ is empty.

The construction of the desired change of basis isomorphism Λ for $H^G_*(B; A)$ is now an obvious generalization of the process presented in Example 4.3. Every generator of $H^G_*(B; A)$ not in $\mathcal{F}_\partial$ but hit by the boundary map ∂ lies above and/or to the left of a generator in $\mathcal{F}_\partial$. Thus, we can push the boundary map off of the generators not in $\mathcal{F}_\partial$.

PROPOSITION 4.5. *Let B be a G-space whose homology $H^G_*(B; A)$ is free over H_* with even-dimensional space-like generators. Assume that the G-space Y is formed from B by adding a single even-dimensional cell of the form DV. Then there is a H_*-module isomorphism $\Lambda : H^G_*(B; A) \longrightarrow H^G_*(B; A)$ such that:*

(i) *the composite*

$$\widetilde{H}^G_*(S^V; A) \xrightarrow{\partial} H^G_{*-1}(B; A) \xrightarrow{\Lambda} H^G_{*-1}(B; A)$$

factors through the summand of $H^G_*(B; A)$ spanned by the generators of $H^G_*(B; A)$ in $\mathcal{F}_\partial$.

(ii) *the map $\Lambda \circ \partial$ hits every generator in $\mathcal{F}_\partial$ in the sense that the composite of this map with the projection of $H^G_*(B; A)$ onto the summand generated by any element of $\mathcal{F}_\partial$ is nonzero.*

(iii) *Λ is the identity map on all the G/e-generators of $H_*^G(B;A)$ and on those G/G-generators of $H_*^G(B;A)$ not in $\mathcal{F}_\partial$.*

Recall that J is the summand of $H_*^G(B;A)$ spanned by the generators in $\mathcal{F}_\partial$, and Z is the summand spanned by all the other generators. Thus, $H_*^G(B;A)$ decomposes as the direct sum $J \oplus Z$. Using Λ, we can now write our cell-attaching long exact sequence in the form

$$\cdots \longrightarrow J \oplus Z \xrightarrow{\tilde{\chi}} H_*^G(Y;A) \xrightarrow{\psi} \widetilde{H}_*^G(S^V;A) \xrightarrow{(\partial',0)} \Sigma(J \oplus Z) \longrightarrow \cdots.$$

Here, $\tilde{\chi}$ is the composite $\Lambda^{-1} \circ \chi$, and ∂' is the composite $\Lambda \circ \partial$ regarded as a map into ΣJ. This sequence is cluttered by the summand Z of $H_*^G(B;A)$ and its image in $H_*^G(Y;A)$. The function of the quotient Q of $H_*^G(Y;A)$ introduced in the previous section is to eliminate this clutter. Note that, since the image of the adjusted boundary map ∂' lies entirely inside the summand J, the composite $Z \subset H_*^G(B;A) \xrightarrow{\chi} H_*^G(Y;A)$ must be a monomorphism. Recall that Q is just the quotient of $H_*^G(Y;A)$ obtained by killing the image of this composite. The long exact sequence for Q introduced in Section 4.1 is a special case of a general algebraic construction which reappears several times in the proof of Theorem 3.3. Thus, we include the following lemma describing that construction.

LEMMA 4.6. *Let*

$$\cdots \longrightarrow J \oplus Z \xrightarrow{\tilde{\chi}} M \xrightarrow{\psi} N \xrightarrow{(\partial',0)} \Sigma(J \oplus Z) \longrightarrow \cdots$$

be a long exact sequence of H_*-modules. Define the H_*-module Q by the short exact sequence

$$0 \longrightarrow Z \longrightarrow M \xrightarrow{\pi} Q \longrightarrow 0.$$

Observe that the map ψ factors through the projection $\pi : M \longrightarrow Q$ to provide a map $\psi' : Q \longrightarrow N$. Also, let $\chi' : J \longrightarrow Q$ be the composite of π and the restriction of $\tilde{\chi}$ to J. Then

$$\cdots \longrightarrow J \xrightarrow{\chi'} Q \xrightarrow{\psi'} N \xrightarrow{\partial'} \Sigma J \longrightarrow \cdots$$

is a long exact sequence of H_*-modules.

PROOF. This follows easily by chasing the diagram

$$\begin{array}{cccccccccc}
 & & 0 & & 0 & & & & & \\
 & & \downarrow & & \downarrow & & & & & \\
 & & Z & \xrightarrow{=} & Z & & & & & \\
 & & \downarrow{\scriptstyle \iota_2} & & \downarrow & & & & & \\
\cdots & \longrightarrow & J \oplus Z & \xrightarrow{\tilde{\chi}} & M & \xrightarrow{\psi} & N & \xrightarrow{(\partial',0)} & \Sigma(J \oplus Z) & \longrightarrow \cdots \\
 & & \downarrow{\scriptstyle \pi_1} & & \downarrow{\scriptstyle \pi} & & \downarrow{\scriptstyle =} & & \downarrow{\scriptstyle \pi_1} & \\
\cdots & \longrightarrow & J & \xrightarrow[\chi']{} & Q & \xrightarrow[\psi']{} & N & \xrightarrow[\partial']{} & \Sigma J & \longrightarrow \cdots \\
 & & \downarrow & & \downarrow & & & & & \\
 & & 0 & & 0 & & & & &
\end{array}$$

in which the maps in the left column are the obvious inclusion and projection. □

By taking M to be $H_*^G(Y;A)$ and N to be $\widetilde{H}_*^G(S^V;A)$ in the lemma above, we obtain our fundamental long exact sequence

$$\cdots \longrightarrow J \xrightarrow{\chi'} Q \xrightarrow{\psi'} \widetilde{H}_*^G(S^V;A) \xrightarrow{\partial'} \Sigma J \longrightarrow \cdots$$

for Q. Hereafter, the free H_*-module $\widetilde{H}_*^G(S^V;A)$ is usually denoted N for notational compactness. Recall that it has a single generator of type G/G in dimension V.

4.3. Characterizing dimension-shifting long exact sequences

In this section, we assume that the elements $\omega_1, \omega_2, \ldots, \omega_n$ of $RO(G)$ form a ramp of length n, $V \in RO(G)$ bounds this ramp, and the elements $\omega_1', \omega_2', \ldots, \omega_{n+1}'$ of $RO(G)$ form a V-shifted ramp as described in Definition 2.3. Moreover, all of these elements of $RO(G)$ are assumed to be even-dimensional and space-like. We also assume that J is a free H_*-module with n G/G-generators in the dimensions ω_i, J' is a free H_*-module with $n+1$ G/G-generators in the dimensions ω_j', and N is a free H_*-module having one G/G-generator in dimension V. Our goal here is to characterize the maps $\bar{\chi} : J \longrightarrow J'$, $\bar{\psi} : J' \longrightarrow N$, and $\bar{\partial} : N \longrightarrow \Sigma J$ for which the sequence

$$\cdots \longrightarrow J \xrightarrow{\bar{\chi}} J' \xrightarrow{\bar{\psi}} N \xrightarrow{\bar{\partial}} \Sigma J \longrightarrow \cdots$$

is a long exact sequence. We refer to such a long exact sequence as a dimension-shifting long exact sequence because of the close connection between such sequences and the cell-attaching long exact sequences arising in situations like Example 2.2. Recall the notions of a standard shift map and of a map constructed from standard shift maps from Definition 1.17.

PROPOSITION 4.7. *The sequence*

$$\cdots \longrightarrow J \xrightarrow{\bar{\chi}} J' \xrightarrow{\bar{\psi}} N \xrightarrow{\bar{\partial}} \Sigma J \longrightarrow \cdots$$

is a long exact sequence if and only if the following four conditions are satisfied:

(i) *$\bar{\chi}$ and $\bar{\psi}$ are constructed from standard shift maps,*

(ii) *each of the components $\bar{\partial}_i : N \longrightarrow \Sigma^{\omega_i+1} H_*$ of the boundary map $\bar{\partial}$ is nonzero,*

(iii) *for $1 \le i \le n$, the composite*

$$J_{\omega_i} \xrightarrow{\bar{\chi}_{\omega_i}} J'_{\omega_i} \xrightarrow{\bar{\psi}_{\omega_i}} N_{\omega_i}$$

is zero, and

(iv) *the composite*

$$N_V \xrightarrow{\bar{\partial}_V} (\Sigma J)_V \xrightarrow{(\Sigma\bar{\chi})_V} (\Sigma J')_V$$

is zero.

This result is proven in Chapter 6. However, by examining the putative long exact sequence in the dimensions of the generators of J, J', and N, it is relatively easy to verify that all four conditions are necessary.

REMARK 4.8. Since J is a free H_*-module with one generator in dimension ω_i, for $1 \le i \le n$, condition (iii) is equivalent to the assertion that $\bar{\psi} \circ \bar{\chi} = 0$. Similarly, because N has one generator in dimension V, condition (iv) is equivalent to the assertion that $\Sigma\bar{\chi} \circ \bar{\partial} = 0$.

It is natural to wonder how hard it is to find maps $\bar{\chi}$, $\bar{\psi}$, and $\bar{\partial}$ satisfying the conditions in this proposition. In the remainder of this section, we show that there is only one obstruction to their existence. The first two of the four conditions in the proposition are quite straightforward, and there are obviously maps $\bar{\chi}$, $\bar{\psi}$, and $\bar{\partial}$ satisfying them. The last two conditions are actually much simplier and more easily satisfied than their appearance suggests. Note that the map $\bar{\chi}$ is completely determined by its behavior in the dimensions ω_i of the generators of J. The only two generators of J' which make nonzero contributions to J' in dimension ω_i are the two in dimensions ω_i' and ω_{i+1}'. Thus, if $j \neq i, i+1$, then the component $\bar{\chi}^{i,j}$ of $\bar{\chi}$ associated to the generators of J and J' in dimensions ω_i and ω_j', respectively, is zero. The composite in condition (iii) therefore has the form

$$A \xrightarrow{\bar{\chi}_{\omega_i}} R \oplus \langle \mathbb{Z} \rangle \xrightarrow{\bar{\psi}_{\omega_i}} \langle \mathbb{Z}/p \rangle$$

with the R coming from the generator of J' in dimension ω_i and the $\langle \mathbb{Z} \rangle$ coming from the generator in dimension ω_{i+1}'. Since $\bar{\chi}$ and $\bar{\psi}$ must be constructed from standard shift maps by condition (i) of the proposition, the composite $\bar{\psi}_{\omega_i} \circ \bar{\chi}_{\omega_i}$ is easily computing by using the multiplicative structure of H_* (see Proposition 1.10).

The computation of $\bar{\psi}_{\omega_i} \circ \bar{\chi}_{\omega_i}$, which we carry out in Section 6.5, reveals that, unless a nontrivial constraint on the dimensions of the generators of J, J', and N is satisfied, there are no maps $\bar{\chi}$ and $\bar{\psi}$ satisfying conditions (i) and (iii). To understand this constraint, recall the function $d : RO_0(G) \longrightarrow \mathbb{Z}$ introduced in Definition 1.4, and note that

$$V + \sum_{1 \le i \le n} \omega_i - \sum_{1 \le j \le n+1} \omega_j'$$

is in $RO_0(G)$.

PROPOSITION 4.9. *There exist maps $\bar{\chi} : J \longrightarrow J'$ and $\bar{\psi} : J' \longrightarrow N$, constructed from standard shift maps, such that $\bar{\psi} \circ \bar{\chi} = 0$ if and only if*

$$d_{(V + \sum \omega_i - \sum \omega_j')} \equiv \pm 1 \mod p.$$

It is easier to obtain maps satisfying condition (iv) of Proposition 4.7.

LEMMA 4.10. *Let $\bar{\chi} : J \longrightarrow J'$ be a map constructed from standard shift maps. Then there exist nonzero maps $\bar{\partial} : N \longrightarrow \Sigma J$ such that the composite*

$$N \xrightarrow{\bar{\partial}} \Sigma J \xrightarrow{\Sigma \bar{\chi}} \Sigma J'$$

is zero. Moreover, each component of any such map $\bar{\partial}$ is nonzero.

PROOF. The composite in condition (iv) of Proposition 4.7 has the form

$$A \xrightarrow{\bar{\partial}_V} \bigoplus_{1 \le i \le n} \langle \mathbb{Z}/p \rangle \xrightarrow{(\Sigma \bar{\chi})_V} \bigoplus_{1 < j \le n} \langle \mathbb{Z}/p \rangle.$$

Each generator of J contributes a copy of $\langle \mathbb{Z}/p \rangle$ to the left direct sum above, and each of the generators of J' except those in dimensions ω_1' and ω_{n+1}' contributes a copy of $\langle \mathbb{Z}/p \rangle$ to the right direct sum. To verify that this composite is zero, it suffices to check that its composite with the projection onto each of the summands of J_V' vanishes. The composite of $(\Sigma \bar{\chi})_V \circ \bar{\partial}_V$ with the projection onto the jth-summand has the form

$$A \longrightarrow \langle \mathbb{Z}/p \rangle \oplus \langle \mathbb{Z}/p \rangle \longrightarrow \langle \mathbb{Z}/p \rangle.$$

Here, the two copies of $\langle \mathbb{Z}/p \rangle$ in the middle come from the generators of J in dimensions ω_{j-1} and ω_j. Both components of the second map are nonzero since $\bar{\chi}$ is constructed from standard shift maps. It follows that the map $(\Sigma\bar{\chi})_V$ is surjective, and so has kernel $\langle \mathbb{Z}/p \rangle$. Moreover, if x is a nonzero element of this kernel, then all n of its coordinates are nonzero. By Lemma 1.12(a), there is a one-to-one correspondence between such nonzero elements and maps $\bar{\partial} : N \longrightarrow \Sigma J$ such that each component of $\bar{\partial}$ is nonzero and $(\Sigma\bar{\chi})_V \circ \bar{\partial}_V = 0$. □

Combining this lemma with Propositions 4.7 and 4.9 yields:

COROLLARY 4.11. *There are maps $\bar{\chi}$, $\bar{\psi}$, and $\bar{\partial}$ for which*

$$\cdots \longrightarrow J \xrightarrow{\bar{\chi}} J' \xrightarrow{\bar{\psi}} N \xrightarrow{\bar{\partial}} \Sigma J \longrightarrow \cdots$$

is a long exact sequence if and only if

$$d_{(V+\sum \omega_i - \sum \omega'_j)} \equiv \pm 1 \mod p.$$

4.4. Constructing the comparison dimension-shifting sequence

We return now to the assumptions about B, Y, and ∂ stated at the beginning of Section 4.1. In Section 4.2, the proof of Theorem 3.3 is reduced to analyzing a long exact sequence of the form

$$\cdots \longrightarrow J \xrightarrow{\chi'} Q \xrightarrow{\psi'} N \xrightarrow{\partial'} \Sigma J \longrightarrow \cdots \tag{4.1}$$

in which $N = \widetilde{H}^G_*(S^V; A)$, J is the summand of $H^G_*(B; A)$ spanned by the generators in the finite set $\mathcal{F}_\partial$, and Q is the quotient of $H^G_*(Y; A)$ by the image of the summand Z of $H^G_*(B; A)$ spanned by the generators not in $\mathcal{F}_\partial$. Recall that the generators of $H^G_*(B; A)$ in $\mathcal{F}_\partial$ lie in dimensions $\omega_1, \omega_2, \ldots, \omega_n$ which form a ramp of length n. To complete the proof of Theorem 3.3, we must show that Q is a free H_*-module having $n+1$ generators in dimensions $\omega'_1, \omega'_2, \ldots, \omega'_{n+1}$ which form a V-shifted ramp associated the the ramp formed by the ω_i.

Our first task is to verify that there are dimensions ω'_i satisfying these conditions in which Q is sufficiently well-behaved to permit the construction an appropriate map comparing it to the free H_*-module $J' = \sum_{j=1}^{n+1} \Sigma^{\omega'_i} H_*$. In order to show that this map is an isomorphism, we must then describe the behavior of long exact sequence (4.1) in the dimensions of the generators of J, J', and N precisely enough to permit the construction of a comparison sequence whose exactness can be proven via Proposition 4.7. These tasks are carried out in the next three propositions and their corollaries. The proofs of two of those propositions are lengthy and are therefore presented separately in Chapter 5.

PROPOSITION 4.12. *There exist space-like even-dimensional elements ω'_1, ω'_2, $\ldots$, ω'_{n+1} of $RO(G)$ forming a V-shifted ramp such that, for $1 \le i \le n+1$, $Q_{\omega'_i} \cong A$. Moreover, if $1 < i < n+1$, then long exact sequence* (4.1) *reduces to the short exact sequence*

$$0 \longrightarrow \langle \mathbb{Z} \rangle \oplus L \xrightarrow{\chi'_{\omega'_i}} A \xrightarrow{\psi'_{\omega'_i}} \langle \mathbb{Z}/p \rangle \longrightarrow 0$$

in dimension ω'_i. For $i = 1$ or $n+1$, this long exact sequence reduces to the short exact sequences

$$0 \longrightarrow L \xrightarrow{\chi'_{\omega'_1}} A \xrightarrow{\psi'_{\omega'_1}} \langle \mathbb{Z} \rangle \longrightarrow 0$$

and

$$0 \longrightarrow \langle \mathbb{Z} \rangle \xrightarrow{\chi'_{\omega'_{n+1}}} A \xrightarrow{\psi'_{\omega'_{n+1}}} R \longrightarrow 0,$$

respectively, in dimension ω'_i. *In these short exact sequences, the copies of* $\langle \mathbb{Z} \rangle$ *and* L *in the left-hand term are contributed by the generators of* J *in dimensions* ω_{i-1} *and* ω_i, *respectively.*

Let $J' = \bigoplus_{1 \le i \le n+1} \Sigma^{\omega'_i} H_*$ be a free H_*-module on generators in the dimensions ω'_i provided by this proposition. We wish to construct a map $\theta : J' \longrightarrow Q$ making a diagram of the form

$$\begin{array}{ccccccccc} \cdots \longrightarrow & J & \xrightarrow{\bar{\chi}} & J' & \xrightarrow{\bar{\psi}} & N & \xrightarrow{\bar{\partial}} & \Sigma J & \longrightarrow \cdots \\ & \downarrow{=} & & \downarrow{\theta} & & \downarrow{=} & & \downarrow{=} & \\ \cdots \longrightarrow & J & \xrightarrow[\chi']{} & Q & \xrightarrow[\psi']{} & N & \xrightarrow[\partial']{} & \Sigma J & \longrightarrow \cdots \end{array} \tag{4.2}$$

commute. Note that the bottom row of this diagram is long exact sequence (4.1). To construct θ, it suffices to specify that map on each of the generators of J'. It is easy to see that J', like Q, is isomorphic to A in the dimensions of those generators. Thus, we can define the desired comparision map θ by taking it to be the identity map of A in dimension ω'_i, for $1 \le i \le n+1$.

To complete this diagram, we must select the maps $\bar{\chi}$ and $\bar{\psi}$. Since we wish to employ Proposition 4.7 to establish the exactness of the top row of this diagram, these maps must be constructed from standard shift maps. The map $\bar{\psi}$ ought to be the composite $\psi' \circ \theta$. The short exact sequences in Proposition 4.12 imply that, if $\bar{\psi}$ is defined in this way, then it has the proper form.

COROLLARY 4.13. *The map* $\bar{\psi} = \psi' \circ \theta$ *is constructed from standard shift maps.*

The map $\bar{\chi} : J \longrightarrow J'$ is obtained by lifting χ' along θ. To show that this lifting exists, we must analyze long exact sequence (4.1) in the dimensions of the generators of J.

PROPOSITION 4.14. *For* $1 \le i \le n$, *long exact sequence* (4.1) *reduces to the short exact sequence*

$$0 \longrightarrow A \xrightarrow{\chi'_{\omega_i}} R \oplus \langle \mathbb{Z} \rangle \xrightarrow{\psi'_{\omega_i}} \langle \mathbb{Z}/p \rangle \longrightarrow 0$$

in dimension ω_i. *Moreover, in these dimensions, the map* $\theta : J' \longrightarrow Q$ *is an isomorphism* $\theta_{\omega_i} : R \oplus \langle \mathbb{Z} \rangle \xrightarrow{\cong} R \oplus \langle \mathbb{Z} \rangle$. *The copies of* R *and* $\langle \mathbb{Z} \rangle$ *in the domain of* θ_{ω_i} *are contributed by the generators of* J' *in dimensions* ω'_i *and* ω'_{i+1}, *respectively.*

The desired lifting $\bar{\chi} : J \longrightarrow J'$ can be defined by assigning it the value $\theta_{\omega_i}^{-1} \circ \chi'_{\omega_i}$ in dimension ω_i for $1 \le i \le n$. Lemma 12.1, which characterizes short exact sequences of the form appearing in the proposition, indicates that χ'_{ω_i} takes the generator μ of $A(G/G)$ to $(\pm\xi, \pm\epsilon)$, where ξ and ϵ are the standard generators of $R(G/G)$ and $\langle \mathbb{Z} \rangle(G/G)$, respectively. These observations suffice for the proof of the following corollary:

COROLLARY 4.15. *There is a map $\bar{\chi} : J \longrightarrow J'$ constructed from standard shift maps which makes the diagram*

$$\begin{array}{ccc} & & J' \\ & \overset{\bar{\chi}}{\nearrow} & \big\downarrow \theta \\ J & \xrightarrow[\chi']{} & Q \end{array}$$

commute.

In order to show that the top row of diagram (4.2) satisfies condition (iv) of Proposition 4.7, we need to understand the behavior of the map θ in dimension $V - 1$.

PROPOSITION 4.16. *In dimension V, the diagram*

$$\begin{array}{ccccccccc} & & & & & & \Sigma J' & & \\ & & & & & \overset{\Sigma\bar{\chi}}{\nearrow} & \big\downarrow \Sigma\theta & & \\ \cdots \longrightarrow & N & \xrightarrow[\partial']{} & \Sigma J & \xrightarrow[\Sigma\chi']{} & & \Sigma Q & \xrightarrow[\Sigma\psi']{} & \Sigma N \longrightarrow \cdots \end{array}$$

has the form

$$\begin{array}{ccccccc} & & & & \langle \mathbb{Z}/p \rangle^{n-1} & & \\ & & & \overset{\bar{\chi}_{V-1}}{\nearrow} & \big\downarrow \theta_{V-1} & & \\ \cdots \longrightarrow & A & \xrightarrow[\partial'_V]{} \langle \mathbb{Z}/p \rangle^{n} & \xrightarrow[\chi'_{V-1}]{} & \langle \mathbb{Z}/p \rangle^{n-1} & \longrightarrow & 0. \end{array}$$

Thus, θ_{V-1} is an isomorphism.

PROOF. The Mackey functors N_V, $(\Sigma J)_V$, $(\Sigma J')_V$, and $(\Sigma N)_V$ are easily computed from the information about H_* contained in Proposition 1.8. The value of $(\Sigma Q)_V$ and the surjectivity of θ_{V-1} then follow from the exactness of the bottom row. Being a surjective map of finite dimensional vector spaces over $\mathbb{Z}/p$ of the same dimension, $\theta_{V-1}(G/G)$ must be an isomorphism. Since the range and domain of θ_{V-1} vanish at G/e, this completes the proof. □

PROOF OF THEOREM 3.3. Proposition 4.12 enables us to construct a map θ comparing Q to the free H_*-module J' to which it should be isomorphic. That proposition and Corollary 4.15 allow us to construct the maps $\bar{\chi}$ and $\bar{\psi}$ which make diagram (4.2) commute. We wish to use Proposition 4.7 to show that the top row of this diagram is a long exact sequence. Corollaries 4.13 and 4.15 indicate that condition (i) is satisfied. Assertion (ii) in Proposition 4.5 indicates that condition (ii) of Proposition 4.7 is satisfied. Condition (iii) follows immediately from the exactness of the bottom row of our diagram, and condition (iv) follows from that exactness and Proposition 4.16. The exactness of the top row clearly implies that θ is an isomorphism so that Q is a free H_*-module with generators in the appropriate dimensions.

Since Q is a free H_*-module, its defining short exact sequence

$$0 \longrightarrow Z \longrightarrow H_*^G(Y; A) \xrightarrow{\pi} Q \longrightarrow 0.$$

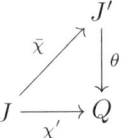

splits, giving an isomorphism

$$H_*^G(Y;A) \cong Z \oplus Q.$$

This implies that $H_*^G(Y;A)$ is a free H_*-module with generators in the specified dimensions. The assertion of the theorem about the behavior of the map $\chi : H_*^G(B;A) \longrightarrow H_*^G(Y;A)$ on the generators of $H_*^G(B;A)$ not in $\mathcal{F}_\partial$ follows from the fact that the isomorphism used to establish the freeness of $H_*^G(Y;A)$ is derived from the inclusion of Z into $H_*^G(Y;A)$.

To verify the claim of the theorem about the behavior of the map χ on the generators of $H_*^G(B;A)$ in $\mathcal{F}_\partial$, we would like to use the diagram

$$\begin{array}{ccc} H_*^G(B;A) & \xrightarrow{\ \chi\ } & H_*^G(Y;A) \\ {\scriptstyle \pi'}\big\downarrow & & \big\downarrow{\scriptstyle \pi} \\ J & \xrightarrow{\ \chi'\ } & Q \end{array}$$

in which π' is the projection onto the summand J of $H_*^G(B;A)$. However, it is not entirely obvious that this diagram commutes. If the map χ were replaced by the map $\tilde{\chi} = \Lambda^{-1} \circ \chi$, then the resulting diagram would certainly commute since it is a part of the appropriate special case of the diagram used to prove Lemma 4.6. Observe that the difference $1 - \Lambda^{-1}$ between the identity map of $H_*^G(B;A)$ and Λ^{-1} factors through Z, basically because the difference between the identity and Λ^{-1} arises from certain elements of Z which are used to adjust the generators of $H_*^G(B;A)$ indexed on $\mathcal{F}_\partial$. It follows that $\pi \circ \tilde{\chi} = \pi \circ \chi$, so the desired diagram does, in fact, commute. The claim about the behavior of the map χ on the generators in $\mathcal{F}_\partial$ can now be checked by examining this diagram in dimension ω_i, for $1 \leq i \leq n$, and applying Proposition 4.14. $\square$

CHAPTER 5

Computing $H_*^G(B \cup DV; A)$ in the key dimensions

Throughout this chapter, B is a G-space whose homology $H_*^G(B;A)$ is free over H_* with even-dimensional space-like generators, and the G-space Y is formed from B by adding a single even-dimensional cell of the form DV. We assume that the boundary map in the associated cell-attaching long exact sequence is nonzero. Recall the quotient long exact sequence (4.1) of that cell-attaching sequence displayed at the beginning of Section 4.4. Our goal here is to prove Propositions 4.12 and 4.14, which describe that quotient long exact sequence in certain critical dimensions. To prove these results, we work with the long exact sequences in homology coming from the short exact sequence

$$0 \longrightarrow L \xrightarrow{f} A \xrightarrow{g} \langle \mathbb{Z} \rangle \longrightarrow 0.$$

of coefficient Mackey functors.

Coupling these long exact sequences with the cell-attaching long exact sequences, we obtain the commuting diagram

$$\begin{array}{ccccccccc}
& & \vdots & & \vdots & & \vdots & & \vdots \\
& & \downarrow & & \downarrow & & \downarrow & & \downarrow \\
\cdots \longrightarrow & & H_*^G(B;L) & \xrightarrow{\chi^L} & H_*^G(Y;L) & \xrightarrow{\psi^L} & \widetilde{H}_*^G(S^V;L) & \xrightarrow{\partial^L} & H_{*-1}^G(B;L) \longrightarrow \cdots \\
& & \downarrow & & \downarrow & & \downarrow & & \downarrow \\
\cdots \longrightarrow & & H_*^G(B;A) & \xrightarrow{\chi} & H_*^G(Y;A) & \xrightarrow{\psi} & \widetilde{H}_*^G(S^V;A) & \xrightarrow{\partial} & H_{*-1}^G(B;A) \longrightarrow \cdots \\
& & \downarrow & & \downarrow & & \downarrow & & \downarrow \\
\cdots \longrightarrow & & H_*^G(B;\langle\mathbb{Z}\rangle) & \xrightarrow{\chi^{\langle\mathbb{Z}\rangle}} & H_*^G(Y;\langle\mathbb{Z}\rangle) & \xrightarrow{\psi^{\langle\mathbb{Z}\rangle}} & \widetilde{H}_*^G(S^V;\langle\mathbb{Z}\rangle) & \xrightarrow{\partial^{\langle\mathbb{Z}\rangle}} & H_{*-1}^G(B;\langle\mathbb{Z}\rangle) \longrightarrow \cdots \\
& & \downarrow & & \downarrow & & \downarrow & & \downarrow \\
& & \vdots & & \vdots & & \vdots & & \vdots
\end{array} \tag{5.1}$$

with exact rows and columns. However, we do not want to work directly with this diagram. Instead, we want to work with a quotient of this diagram obtained by killing off everything associated to the summand Z of $H_*^G(B;A)$. One row of this quotient diagram is the quotient long exact sequence (4.1) and the other two rows are the analogous long exact sequences for L and $\langle \mathbb{Z} \rangle$ coefficients. Our first objective is to define this quotient diagram and show that it has exact rows and columns. This is done in the first section below. This quotient diagram is then used to prove Propositions 4.12 and 4.14 in the second section.

5.1. Using the Universal Coefficient Theorem

Recall the finite subset $\mathcal{F}_\partial$ of the set of G/G-generators of $H_*^G(B;A)$ selected in Definition 4.4. From $\mathcal{F}_\partial$, we obtain the direct sum decompositon $H_*^G(B;A) \cong J \oplus Z$

in which J and Z are the summands spanned by the generators in $\mathcal{F}_\partial$ and those not in $\mathcal{F}_\partial$, respectively. Recall also the map $\partial' : \widetilde{H}_*^G(S^V; A) \longrightarrow \Sigma J$ and automorphism Λ of $H_*^G(B; A)$ introduced in Proposition 4.5. Denote the composite of Λ and the direct sum decomposition by $\Phi : J \oplus Z \longrightarrow H_*^G(B; A)$. The commuting diagram

$$\begin{array}{ccc} & \widetilde{H}_*^G(S^V; A) & \\ {\scriptstyle (\partial',0)} \swarrow & & \searrow {\scriptstyle \partial} \\ \Sigma(J \oplus Z) & \xrightarrow[\Sigma\Phi]{} & \Sigma H_*^G(B; A) \end{array}$$

describes the connection between the boundary map ∂ of our cell-attaching long exact sequence and the maps ∂' and Φ.

For any Mackey functor S, denote the $RO(G)$-graded homology of a point with S-coefficients by H_*^S. Let $J^S = J \,\square_{H_*}\, H_*^S$ and $Z^S = Z \,\square_{H_*}\, H_*^S$. Since $H_*^G(B; A)$ and $\widetilde{H}_*^G(S^V; A)$ are free H_*-modules, the edge homomorphisms

$$\sigma_B^S : H_*^G(B; A) \,\square_{H_*}\, H_*^S \longrightarrow H_*^G(B; S)$$

and

$$\sigma_V^S : \widetilde{H}_*^G(S^V; A) \,\square_{H_*}\, H_*^S \longrightarrow \widetilde{H}_*^G(S^V; S)$$

of the universal coefficient spectral sequence are isomorphisms (see Proposition 11.1 for the cases which matter here). Let Φ^S be the composite isomorphism

$$J^S \oplus Z^S \cong (J \oplus Z) \,\square_{H_*}\, H_*^S \xrightarrow{\Phi \square_{H_*} 1} H_*^G(B; A) \,\square_{H_*}\, H_*^S \xrightarrow{\sigma_B^S} H_*^G(B; S).$$

Also, let $\partial'_S : \widetilde{H}_*^G(S^V; S) \longrightarrow \Sigma J^S$ be the composite

$$\widetilde{H}_*^G(S^V; S) \xrightarrow{(\sigma_V^S)^{-1}} \widetilde{H}_*^G(S^V; A) \,\square_{H_*}\, H_*^S \xrightarrow{\partial' \square_{H_*} 1} \Sigma J \,\square_{H_*}\, H_*^S = \Sigma J^S.$$

The naturality of the edge homomorphism implies that the diagram

$$\begin{array}{ccc} & \widetilde{H}_*^G(S^V; S) & \\ {\scriptstyle (\partial'_S,0)} \swarrow & & \searrow {\scriptstyle \partial^S} \\ \Sigma(J^S \oplus Z^S) & \xrightarrow[\Sigma\Phi^S]{} & \Sigma H_*^G(B; S) \end{array}$$

commutes. Using this diagram, we can write the homology cell-attaching long exact sequence for Y with S coefficients as

$$\cdots \longrightarrow J^S \oplus Z^S \xrightarrow{\tilde{\chi}^S} H_*^G(Y; S) \xrightarrow{\psi^S} \widetilde{H}_*^G(S^V; S) \xrightarrow{(\partial'_S,0)} \Sigma(J^S \oplus Z^S) \longrightarrow \cdots. \quad (5.2)$$

Here, $\tilde{\chi}^S = \chi^S \circ \Phi^S$. This long exact sequence is natural in S if we define the maps $J^S \longrightarrow J^{S'}$ and $Z^S \longrightarrow Z^{S'}$ associated to a coefficient map $S \longrightarrow S'$ in the obvious way.

The long exact sequence above implies that the restriction $Z^S \longrightarrow H_*^G(Y; S)$ of $\tilde{\chi}^S$ to Z^S is a monomorphism. Define Q^S by the short exact sequence

$$0 \longrightarrow Z^S \longrightarrow H_*^G(Y; S) \xrightarrow{\pi^S} Q^S \longrightarrow 0.$$

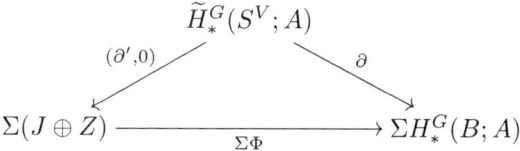

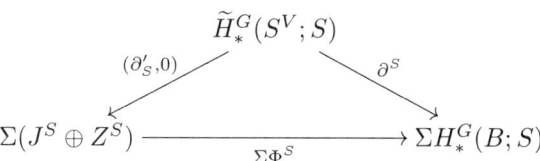

Note that this construction is natural in S since the map $\tilde{\chi}^S$ is natural in S. By applying Lemma 4.6 to long exact sequence (5.2), we obtain the long exact sequence

$$\cdots \longrightarrow J^S \xrightarrow{\chi'_S} Q^S \xrightarrow{\psi'_S} \widetilde{H}_*^G(S^V; S) \xrightarrow{\partial'_S} \Sigma J^S \longrightarrow \cdots,$$

which is also natural in S. Hereafter, we denote $\widetilde{H}_*^G(S^V; S)$ by N^S for consistency with our other notation. The naturality of this sequence allows us to construct the desired quotient of diagram (5.1). However, we need one further result to ensure that the columns of the resulting diagram are exact.

LEMMA 5.1. *There are maps $Q^L \longrightarrow Q$, $Q \longrightarrow Q^{\langle \mathbb{Z} \rangle}$, and $Q^{\langle \mathbb{Z} \rangle} \longrightarrow \Sigma Q^L$ such that the diagram*

$$\begin{array}{ccccccccc}
\cdots \longrightarrow & H_*^G(Y; L) & \longrightarrow & H_*^G(Y; A) & \longrightarrow & H_*^G(Y; \langle \mathbb{Z} \rangle) & \longrightarrow & \Sigma H_*^G(Y; L) & \longrightarrow \cdots \\
& \downarrow{\scriptstyle \pi^L} & & \downarrow{\scriptstyle \pi} & & \downarrow{\scriptstyle \pi^{\langle \mathbb{Z} \rangle}} & & \downarrow{\scriptstyle \Sigma \pi^L} & \\
\cdots \longrightarrow & Q^L & \longrightarrow & Q & \longrightarrow & Q^{\langle \mathbb{Z} \rangle} & \longrightarrow & \Sigma Q^L & \longrightarrow \cdots
\end{array}$$

commutes and has an exact bottom row.

PROOF. Consider the diagram

$$\begin{array}{ccccccccc}
& 0 & & 0 & & 0 & & 0 & \\
& \downarrow & & \downarrow & & \downarrow & & \downarrow & \\
\cdots \longrightarrow & Z^L & \longrightarrow & Z & \longrightarrow & Z^{\langle \mathbb{Z} \rangle} & \longrightarrow & \Sigma Z^L & \longrightarrow \cdots \\
& \downarrow & & \downarrow & & \downarrow & & \downarrow & \\
\cdots \longrightarrow & H_*^G(Y; L) & \longrightarrow & H_*^G(Y; A) & \longrightarrow & H_*^G(Y; \langle \mathbb{Z} \rangle) & \longrightarrow & \Sigma H_*^G(Y; L) & \longrightarrow \cdots \\
& \downarrow{\scriptstyle \pi^L} & & \downarrow{\scriptstyle \pi} & & \downarrow{\scriptstyle \pi^{\langle \mathbb{Z} \rangle}} & & \downarrow{\scriptstyle \Sigma \pi^L} & \\
\cdots \dashrightarrow & Q^L & \dashrightarrow & Q & \dashrightarrow & Q^{\langle \mathbb{Z} \rangle} & \dashrightarrow & \Sigma Q^L & \dashrightarrow \cdots \\
& \downarrow & & \downarrow & & \downarrow & & \downarrow & \\
& 0 & & 0 & & 0 & & 0 &
\end{array}$$

in which the columns are exact. The top half of this diagram commutes because of the naturality of the inclusion $Z^S \longrightarrow H_*^G(Y; S)$ with respect to S. Moreover, the top row of this diagram is exact since it is obtained by taking the box product of the long exact sequence

$$\cdots \longrightarrow H_*^L \longrightarrow H_* \longrightarrow H_*^{\langle \mathbb{Z} \rangle} \longrightarrow \Sigma H_*^L \longrightarrow \cdots \tag{5.3}$$

with the free H_*-module Z. It follows that there are unique choices for the dotted arrows on the bottom row which make the whole diagram commute. A straightforward diagram chase then gives that the bottom row is exact. □

PROPOSITION 5.2. *The diagram*

$$
\begin{array}{ccccccccccc}
 & & \vdots & & \vdots & & \vdots & & \vdots & & \\
 & & \downarrow & & \downarrow & & \downarrow & & \downarrow & & \\
\cdots & \longrightarrow & J^L & \xrightarrow{\chi'_L} & Q^L & \xrightarrow{\psi'_L} & N^L & \xrightarrow{\partial'_L} & \Sigma J^L & \longrightarrow & \cdots \\
 & & \downarrow & & \downarrow & & \downarrow & & \downarrow & & \\
\cdots & \longrightarrow & J & \xrightarrow{\chi'} & Q & \xrightarrow{\psi'} & N & \xrightarrow{\partial'} & \Sigma J & \longrightarrow & \cdots \\
 & & \downarrow & & \downarrow & & \downarrow & & \downarrow & & \\
\cdots & \longrightarrow & J\langle\mathbb{Z}\rangle & \xrightarrow{\chi'_{\langle\mathbb{Z}\rangle}} & Q\langle\mathbb{Z}\rangle & \xrightarrow{\psi'_{\langle\mathbb{Z}\rangle}} & N\langle\mathbb{Z}\rangle & \xrightarrow{\partial'_{\langle\mathbb{Z}\rangle}} & \Sigma J\langle\mathbb{Z}\rangle & \longrightarrow & \cdots, \\
 & & \downarrow & & \downarrow & & \downarrow & & \downarrow & & \\
 & & \vdots & & \vdots & & \vdots & & \vdots & &
\end{array}
\qquad (5.4)
$$

obtained from diagram (5.1) by collapsing out everything associated to Z, commutes and has exact rows and columns.

PROOF. The exactness of the rows in the diagram follows from Lemma 4.6, and the commutativity of the diagram follows from the naturality of the construction described in that lemma. The Q column of the diagram is exact by Lemma 5.1. The N column is just a long exact coefficient sequence for the space S^V. The J column is exact because it is obtained by taking the box product over H_* of long exact sequence (5.3) with a free H_*-module. □

5.2. Constructing the maps of the comparison sequence

In this section, we prove Propositions 4.12 and 4.14, which describe the cell-attaching long exact sequence for the G-space Y in certain critical dimensions. Our basic tool in these proofs is diagram (5.4) of Proposition 5.2. Perhaps the most delicate part of these proofs is selecting the elements ω'_i of $RO(G)$. We have already indicated where these elements ought to appear in our standard plot of elements of $RO(G)$. However, for $p \geq 5$, more than one element of $RO(G)$ plots to each of these locations. Thus, in the early stages of our argument, we look at an arbitrary element ω of $RO(G)$ which plots to one of these locations. Once we have learned enough about the appearance of diagram (5.4) in such a dimension ω, we can then make the appropriate choice for each of the ω'_i.

The first step in these proofs is analyzing the map $\partial'_L : N^L \longrightarrow \Sigma J^L$ in certain dimensions.

LEMMA 5.3. *Let ω be an element of $RO(G)$ such that either*

(i) $|\omega| = |V|$ *and* $|\omega^G| = |\omega_n^G|$

or

(ii) $|\omega| = |\omega_i|$ *and* $|\omega^G| = |\omega_{i-1}^G|$ *for some i such that* $1 < i \leq n$.

Then the map $(\partial'_L)_\omega : N^L_\omega \longrightarrow (\Sigma J^L)_\omega$ is nonzero.

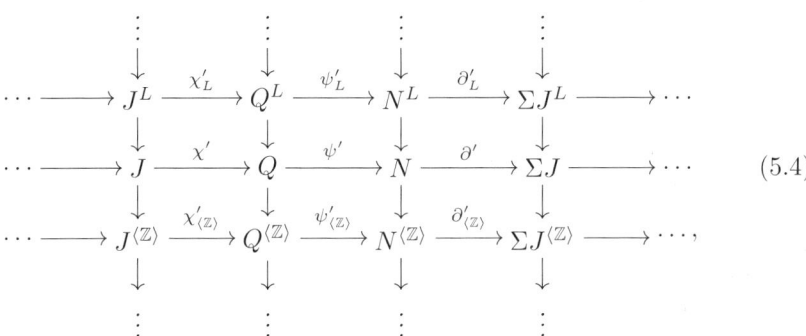

PROOF. Recall that N^L and ΣJ^L are obtained from free H_*-modules by taking a box product over H_* with H_*^L. By Corollary 9.3, the H_*-modules H_*^L and $\Sigma^{2-\xi}H_*^R$ are isomorphic for any nontrivial irreducible G-representation ξ. Thus, $N^L \cong \Sigma^{2-\xi}N^R$ and $\Sigma J^L \cong \Sigma^{3-\xi}J^R$. Further, since the maps ∂'_L and ∂'_R are obtained from ∂' by taking a box product over H_* with H_*^L and H_*^R, respectively, ∂'_L is identified with $\Sigma^{2-\xi}\partial'_R$ under these two isomorphisms. Thus, it suffices to prove that the map $\Sigma^{2-\xi}\partial'_R : \Sigma^{2-\xi}N^R \longrightarrow \Sigma^{3-\xi}J^R$ is nonzero in the indicated dimensions. This task is simplified by the fact that H_*^R, N^R, and J^R are quotients of H_*, N, and J, respectively. By assumption, each component of the map $\partial' : N \longrightarrow J$ is nonzero in dimension V. It follows easily that each component of the map $\Sigma^{2-\xi}\partial'_R : \Sigma^{2-\xi}N^R \longrightarrow \Sigma^{3-\xi}J^R$ is nonzero in dimension $V + 2 - \xi$. Now assume that $\omega \in RO(G)$ satisfies one of the two conditions in the lemma. Then $|\omega| \leq |V + 2 - \xi|$ and $|\omega^G| \geq |(V + 2 - \xi)^G|$. Further, the case $|\omega| = |V + 2 - \xi|$ and $|\omega^G| = |(V + 2 - \xi)^G|$ can occur only if $n = 1$.

Consider first this special case of two equalities. In this case, $\Sigma^{3-\xi}J^R$ is $\langle \mathbb{Z}/p \rangle$ in dimensions $V + 2 - \xi$ and ω. The map $\Sigma^{2-\xi}\partial'_R$ must then be surjective in dimension $V + 2 - \xi$ since it is nonzero. From this and Corollary 8.13, it follows that the map $\Sigma^{2-\xi}\partial'_R$ is surjective, and therefore nonzero, in dimension ω.

Hereafter, we can assume that at least one of the two inequalities relating $V + 2 - \xi$ and ω is strict. In this case, $H_{\omega-(V+2-\xi)}$ is one of the Mackey functors R, $\langle \mathbb{Z} \rangle$, or $\langle \mathbb{Z}/p \rangle$, and so is generated at G/G by an element of the form ξ, ϵ, or $\epsilon\xi$, respectively. Note that there is an integer j such that $1 \leq j \leq n$ and $|\omega^G| = |\omega_j^G|$. In dimension ω, $\Sigma^{3-\xi}J^R$ consists of a single copy of $\langle \mathbb{Z}/p \rangle$ contributed by the generator of J in dimension ω_j. Multiplication by the generator of $H_{\omega-(V+2-\xi)}(G/G)$ induces an isomorphism from $(\Sigma^{3-\xi}J_j^R)_{V+2-\xi}$ to $(\Sigma^{3-\xi}J_j^R)_\omega = (\Sigma^{3-\xi}J^R)_\omega$. Since the jth component $(\Sigma^{2-\xi}\partial'_R)_j : \Sigma^{2-\xi}N^R \longrightarrow \Sigma^{3-\xi}J_j^R$ of the map $\Sigma^{2-\xi}\partial'_R$ is an H_*-module map and is nonzero in dimension $V + 2 - \xi$, it follows that this map is nonzero in dimension ω. $\square$

Observe that an element ω satisfying condition (i) in this lemma plots to the location at which the element ω'_{n+1} should plot. In fact, this lemma provides us with enough information about such an element ω to establish the existence of an element ω'_{n+1} of $RO(G)$ with all of the appropriate properties.

PROPOSITION 5.4. *There exists an element ω'_{n+1} of $RO(G)$ such that*

(i) $|\omega'_{n+1}| = |V|$ *and* $|(\omega'_{n+1})^G| = |\omega_n{}^G|$,

(ii) $Q_{\omega'_{n+1}} \cong A$, *and*

(iii) *in dimension ω'_{n+1}, the middle row in diagram (5.4) is a short exact sequence of the form*

$$0 \longrightarrow \langle \mathbb{Z} \rangle \xrightarrow{\chi'_{\omega'_{n+1}}} A \xrightarrow{\psi'_{\omega'_{n+1}}} R \longrightarrow 0.$$

PROOF. Let ω be any element of $RO(G)$ such that $|\omega| = |V|$ and $|\omega^G| = |\omega_n^G|$. It follows easily from the description of $H_*^{\langle \mathbb{Z} \rangle}$ and H_*^L given in Propositions 9.1, 9.2, and 9.5 that diagram (5.4) has the form

$$\begin{array}{ccccccccc}
 & & & & 0 & & & & \\
 & & & & \downarrow & & & & \\
 & & 0 & \xrightarrow{(\chi'_L)_\omega} & Q_\omega^L & \xrightarrow{(\psi'_L)_\omega} & R & \xrightarrow{(\partial'_L)_\omega} & \langle \mathbb{Z}/p \rangle \\
 & & \downarrow & & \downarrow & & \downarrow{\cong} & & \downarrow \\
0 & \longrightarrow & \langle \mathbb{Z} \rangle & \xrightarrow{\chi'_\omega} & Q_\omega & \xrightarrow{\psi'_\omega} & R & \xrightarrow{\partial'_\omega} & 0 \\
 & & \downarrow{\cong} & & \downarrow{\bar{\epsilon}} & & \downarrow & & \\
0 & \longrightarrow & \langle \mathbb{Z} \rangle & \xrightarrow[(\chi'_{\langle \mathbb{Z} \rangle})_\omega]{} & Q_\omega^{\langle \mathbb{Z} \rangle} & \xrightarrow[(\psi'_{\langle \mathbb{Z} \rangle})_\omega]{} & 0 & &
\end{array}$$

in dimension ω. Lemma 5.3 indicates that the map $(\partial'_L)_\omega$ is nonzero. It follows from Lemma 12.3 that $Q_\omega^L \cong L$.

A simple rank argument applied to the Q column of the diagram implies that the map $\bar{\epsilon}$ is nonzero at G/G. Even though this map need not be onto, its image must then be a copy of $\langle \mathbb{Z} \rangle$. We can therefore derive a short exact sequence of the form

$$0 \longrightarrow L \longrightarrow Q_\omega \longrightarrow \langle \mathbb{Z} \rangle \longrightarrow 0$$

from the Q column of the diagram. By Lemma 12.2(c), the only common solution to this short exact sequence and the short exact sequence in the middle row of the diagram above is a Mackey functor of the form $A[d]$, for some integer d prime to p. Corollary 8.15 now indicates that we can select ω such that $Q_\omega \cong A$. Taking ω'_{n+1} to be this ω completes the proof. □

Selecting the elements ω'_i, for $1 \leq i \leq n$, requires a bit more effort and must be done by starting with ω'_n and working inductively downward to ω'_1. Observe that an element ω satisfying condition (ii) in Lemma 5.3 plots to the location at which the element ω'_i should plot. Lemma 5.3 does not allow us to determine Q as completely in a dimension ω satisfying condition (ii) as it does for a dimension satisfying condition (i). However, we can significantly restrict the possible values of Q in a dimension satisfying condition (ii).

PROPOSITION 5.5. *Let i be an integer such that $1 \leq i \leq n$, and let ω be an element of $RO(G)$ such that $|\omega| = |\omega_i|$ and*

$$|\omega^G| = \begin{cases} |\omega^G_{i-1}| & \text{if } i > 1, \\ |V^G| & \text{if } i = 1. \end{cases}$$

Then Q_ω is either $\langle \mathbb{Z} \rangle \oplus L$ or $A[d]$, for some integer d prime to p. Moreover, if $Q_\omega = \langle \mathbb{Z} \rangle \oplus L$, then $Q_{\omega'} = \langle \mathbb{Z} \rangle \oplus L$ for any other $\omega' \in RO(G)$ plotting to the same position as ω. In dimension ω, the middle row of diagram (5.4) is a short exact sequence of the form

$$0 \longrightarrow \langle \mathbb{Z} \rangle \oplus L \xrightarrow{\chi'_\omega} Q_\omega \xrightarrow{\psi'_\omega} \langle \mathbb{Z}/p \rangle \longrightarrow 0 \qquad \text{if } i > 1$$

or

$$0 \longrightarrow L \xrightarrow{\chi'_\omega} Q_\omega \xrightarrow{\psi'_\omega} \langle \mathbb{Z} \rangle \longrightarrow 0 \qquad \text{if } i = 1.$$

PROOF. For $i = 1$, it is easy to see that the middle row of diagram (5.4) must have the indicated form. This, together with Lemma 12.2(a), implies that Q_ω is either $\langle \mathbb{Z} \rangle \oplus L$ or $A[d]$. The assertion about Q_ω being $\langle \mathbb{Z} \rangle \oplus L$ implying that $Q_{\omega'}$ is also $\langle \mathbb{Z} \rangle \oplus L$ follows from Corollary 8.13.

For $i > 1$, observe that diagram (5.4) has the form

$$\begin{array}{ccccccccc}
 & & 0 & & 0 & & & & \\
 & & \downarrow & & \downarrow & & & & \\
0 & \longrightarrow & L & \xrightarrow{(\chi'_L)_\omega} & Q^L_\omega & \xrightarrow{(\psi'_L)_\omega} & \langle \mathbb{Z}/p \rangle & \xrightarrow{(\partial'_L)_\omega} & \langle \mathbb{Z}/p \rangle \\
 & & \downarrow{\scriptstyle i_2} & & \downarrow & & \downarrow{\scriptstyle \cong} & & \downarrow \\
0 & \longrightarrow & \langle \mathbb{Z} \rangle \oplus L & \xrightarrow{\chi'_\omega} & Q_\omega & \xrightarrow{\psi'_\omega} & \langle \mathbb{Z}/p \rangle & \xrightarrow{\partial'_\omega} & 0 \\
 & & \downarrow{\scriptstyle \pi_1} & & \downarrow{\scriptstyle \bar{\epsilon}} & & \downarrow & & \\
0 & \longrightarrow & \langle \mathbb{Z} \rangle & \xrightarrow[(\chi'_{\langle \mathbb{Z} \rangle})_\omega]{} & Q^{\langle \mathbb{Z} \rangle}_\omega & \xrightarrow[(\psi'_{\langle \mathbb{Z} \rangle})_\omega]{} & 0 & &
\end{array}$$

in dimension ω. There are far too many solutions for the extension problem displayed in the middle row of this diagram for this row to be of use in identifying Q_ω. However, the map $(\partial'_L)_\omega$ is nonzero by Lemma 5.3. This map is therefore an isomorphism, and Q^L_ω must be isomorphic to L. As in the proof of Proposition 5.4, we can argue that the image of the map $\bar{\epsilon}$ is a copy of $\langle\mathbb{Z}\rangle$ which may, or may not, be all of $Q^{\langle\mathbb{Z}\rangle}_\omega$. Regardless, we can extract from the Q column of this diagram a short exact sequence of the form

$$0 \longrightarrow L \longrightarrow Q_\omega \longrightarrow \langle\mathbb{Z}\rangle \longrightarrow 0\,.$$

By Lemma 12.2(a), the only possible solutions to this extension problem are $\langle\mathbb{Z}\rangle \oplus L$ and $A[d]$, for some integer d prime to p. The claim in the proposition about Q_ω being $\langle\mathbb{Z}\rangle \oplus L$ implying that $Q_{\omega'}$ is also $\langle\mathbb{Z}\rangle \oplus L$ follows, as in the case $i = 1$, from Corollary 8.13. □

We turn now to a pair of propositions which set the stage for an inductive proof of Propositions 4.12 and 4.14.

PROPOSITION 5.6. *Let i be an integer such that $1 \le i \le n$, and assume that there is an element ω'_{i+1} of $RO(G)$ such that*

(i) $|\omega'_{i+1}| = \begin{cases} |\omega_{i+1}| & \text{if } i < n \\ |V| & \text{if } i = n, \end{cases}$

(ii) $|(\omega'_{i+1})^G| = |\omega_i^G|$, *and*

(iii) $Q_{\omega'_{i+1}} \cong A$.

Then $Q_{\omega_i} \cong R \oplus \langle\mathbb{Z}\rangle$, and the middle row of diagram (5.4) has the form

$$0 \longrightarrow A \xrightarrow{\chi'_{\omega_i}} R \oplus \langle\mathbb{Z}\rangle \xrightarrow{\psi'_{\omega_i}} \langle\mathbb{Z}/p\rangle \longrightarrow 0$$

in dimension ω_i.

PROOF. The middle row of diagram (5.4) clearly has the form

$$0 \longrightarrow A \xrightarrow{\chi'_{\omega_i}} Q_{\omega_i} \xrightarrow{\psi'_{\omega_i}} \langle\mathbb{Z}/p\rangle \longrightarrow 0$$

in dimension ω_i. By Lemma 12.1, the only possible solutions of this extension problem are $A \oplus \langle\mathbb{Z}/p\rangle$, which occurs if the sequence splits, and $R \oplus \langle\mathbb{Z}\rangle$. Assume that this sequence splits so that $Q_{\omega_i} \cong A \oplus \langle\mathbb{Z}/p\rangle$. The Mackey functor $H_{\omega_i - \omega'_{i+1}}$ is isomorphic to $\langle\mathbb{Z}\rangle$, and is generated at G/G by the element $\epsilon_{\omega_i - \omega'_{i+1}}$. Multiplication by $\epsilon_{\omega_i - \omega'_{i+1}}$ gives a map from the middle row of diagram (5.4) in dimension ω'_{i+1} to that row in dimension ω_i.

If $i < n$, this map of short exact sequences has the form

$$\begin{array}{ccccccccc}
0 & \longrightarrow & \langle\mathbb{Z}\rangle \oplus L & \xrightarrow{\chi'_{\omega'_{i+1}}} & A & \xrightarrow{\psi'_{\omega'_{i+1}}} & \langle\mathbb{Z}/p\rangle & \longrightarrow & 0 \\
 & & \downarrow{\scriptstyle \epsilon'} & & \downarrow{\scriptstyle \epsilon} & & \cong\downarrow{\scriptstyle \epsilon''} & & \\
0 & \longrightarrow & A & \xrightarrow{\chi'_{\omega_i}} & A \oplus \langle\mathbb{Z}/p\rangle & \xrightarrow{\psi'_{\omega_i}} & \langle\mathbb{Z}/p\rangle & \longrightarrow & 0.
\end{array}$$

Let κ', μ', and τ' be the usual elements of $Q_{\omega'_{i+1}} \cong A$ at G/G, and let κ be the usual element of $J_{\omega_i} \cong A$ at G/G. Denote by $(1,0)$ the generator of the first summand of $J_{\omega'_{i+1}} \cong \langle\mathbb{Z}\rangle \oplus L$ at G/G. Proposition 1.10(o) indicates that $(\epsilon'(G/G))(1,0) = \kappa$.

The exactness of the top row implies that $(\chi'_{\omega'_{i+1}}(G/G))(1,0) = \pm\kappa'$. Since we have assumed that the bottom row splits, it follows that $(\epsilon\,(G/G))(\kappa') = (\pm\kappa, 0)$ in $(A \oplus \langle \mathbb{Z}/p \rangle)(G/G)$. The map ϵ vanishes at G/e, so $(\epsilon\,(G/G))(\tau') = 0$ and $(\epsilon\,(G/G))(\mu') = (a\kappa, x)$ for some integer a and some $x \in \mathbb{Z}/p$. Recall however that $\kappa' = p\mu' - \tau'$. From this we get the contradiction that $(\pm\kappa, 0) = (pa\kappa, 0)$ in $(A \oplus \langle \mathbb{Z}/p \rangle)(G/G)$. Thus, the bottom row cannot split, and $Q_{\omega_i} \cong R \oplus \langle \mathbb{Z} \rangle$.

If $i = n$, then the map of short exact sequences given by multiplication by $\epsilon_{\omega_n - \omega'_{n+1}}$ has the form

$$\begin{array}{ccccccccc}
0 & \longrightarrow & \langle \mathbb{Z} \rangle & \xrightarrow{\chi'_{\omega'_{n+1}}} & A & \xrightarrow{\psi'_{\omega'_{n+1}}} & R & \longrightarrow & 0 \\
 & & \downarrow{\scriptstyle \epsilon'} & & \downarrow{\scriptstyle \epsilon} & & \downarrow{\scriptstyle \epsilon''} & & \\
0 & \longrightarrow & A & \xrightarrow{\chi'_{\omega_n}} & A \oplus \langle \mathbb{Z}/p \rangle & \xrightarrow{\psi'_{\omega_n}} & \langle \mathbb{Z}/p \rangle & \longrightarrow & 0.
\end{array}$$

Taking κ, κ', μ', and τ' to be as in the previous case, and looking at the image of the generator of $J_{\omega'_{n+1}} \cong \langle \mathbb{Z} \rangle$ at G/G under ϵ' and $\chi'_{\omega'_{n+1}}$, we again obtain that $(\epsilon\,(G/G))(\kappa') = (\pm\kappa, 0) \in (A \oplus \langle \mathbb{Z}/p \rangle)(G/G)$. The map ϵ still vanishes at G/e, so $(\epsilon\,(G/G))(\tau') = 0$ and $(\epsilon\,(G/G))(\mu') = (a\kappa, x)$ for some integer a and some $x \in \mathbb{Z}/p$. Thus, the equation $\kappa' = p\mu' - \tau'$ still gives us the contradiction that $(\pm\kappa, 0) = (pa\kappa, 0)$ in $(A \oplus \langle \mathbb{Z}/p \rangle)(G/G)$. Again, it follows that the bottom row cannot split, so $Q_{\omega_n} \cong R \oplus \langle \mathbb{Z} \rangle$. □

PROPOSITION 5.7. *Let i be an integer such that $1 \leq i \leq n$, and assume that $Q_{\omega_i} \cong R \oplus \langle \mathbb{Z} \rangle$. Then there is an element ω'_i of $RO(G)$ such that*

(i) $|(\omega'_i)^G| = \begin{cases} |\omega^G_{i-1}| & \text{if } i > 1 \\ |V^G| & \text{if } i = 1, \end{cases}$

(ii) $|\omega'_i| = |\omega_i|$, *and*

(iii) $Q_{\omega'_i} \cong A$.

PROOF. Let ω be an element of $RO(G)$ such that $|\omega| = |\omega_i|$, and $|\omega^G|$ is $|\omega^G_{i-1}|$ or $|V^G|$, depending on whether $i > 1$ or not. By Proposition 5.5, we know that Q_ω is either $\langle \mathbb{Z} \rangle \oplus L$ or $A[d]$, for some integer d prime to p. If $Q_\omega = A[d]$, then Corollary 8.15 allows us to pick an element ω'_i of $RO(G)$ satisfying the three conditions in the proposition. Thus, it suffices to eliminate the possibility that $Q_\omega = \langle \mathbb{Z} \rangle \oplus L$. Assume, to the contrary, that Q_ω is $\langle \mathbb{Z} \rangle \oplus L$. Observe that $H_{\omega_i - \omega}$ is the Mackey functor R, which is generated at G/G by the element $\xi_{\omega_i - \omega}$. Multiplication by $\xi_{\omega_i - \omega}$ gives a map from the middle row of diagram (5.4) in dimension ω to that row in dimension ω_i.

If $i > 1$, this map of short exact sequences has the form

$$\begin{array}{ccccccccc}
0 & \longrightarrow & \langle \mathbb{Z} \rangle \oplus L & \xrightarrow{\chi'_{\omega}} & \langle \mathbb{Z} \rangle \oplus L & \xrightarrow{\psi'_{\omega}} & \langle \mathbb{Z}/p \rangle & \longrightarrow & 0 \\
 & & \downarrow{\scriptstyle \xi'} & & \downarrow{\scriptstyle \xi} & & \cong\downarrow{\scriptstyle \xi''} & & \\
0 & \longrightarrow & A & \xrightarrow{\chi'_{\omega_i}} & R \oplus \langle \mathbb{Z} \rangle & \xrightarrow{\psi'_{\omega_i}} & \langle \mathbb{Z}/p \rangle & \longrightarrow & 0.
\end{array}$$

There are no nonzero maps from $\langle \mathbb{Z} \rangle$ to L or from L to $\langle \mathbb{Z} \rangle$. This, plus the fact that the map χ'_ω must be an isomorphism at G/e implies that, by picking orientations correctly, we can assume that the map χ'_ω is $p \oplus id$, where p denotes the multiplication by p map. It follows easily that the restriction of the composite $\xi'' \circ \psi'_\omega$ to the $\langle\mathbb{Z}\rangle$ summand of its domain must be surjective. The map ξ is derived from multiplication by $\xi_{\omega_i - \omega}$, and so is a composite of the form

$$\langle\mathbb{Z}\rangle \oplus L \cong A \,\square\, (\langle\mathbb{Z}\rangle \oplus L) \xrightarrow{\tilde{\xi}\square 1} R \,\square\, (\langle\mathbb{Z}\rangle \oplus L) \longrightarrow R \oplus \langle\mathbb{Z}\rangle,$$

where $\tilde{\xi}$ takes $\mu \in A$ to $\xi_{\omega_i - \omega} \in R$. Table 1.1 gives that $R \,\square\, \langle\mathbb{Z}\rangle \cong \langle\mathbb{Z}/p\rangle$. But there is no p-torsion in $R \oplus \langle\mathbb{Z}\rangle$, so the map ξ must be zero on the summand $\langle\mathbb{Z}\rangle$ of its domain. Since the right square in the diagram commutes, this contradicts the surjectivity of the restriction of $\xi'' \circ \psi'_\omega$ to that summand. It follows that $Q_\omega \neq \langle\mathbb{Z}\rangle \oplus L$ if $i > 1$.

If $i = 1$, then the map of short exact sequences given by multiplication by $\xi_{\omega_1 - \omega}$ has the form

$$\begin{array}{ccccccccc}
0 & \longrightarrow & L & \xrightarrow{\chi'_\omega} & \langle\mathbb{Z}\rangle \oplus L & \xrightarrow{\psi'_\omega} & \langle\mathbb{Z}\rangle & \longrightarrow & 0 \\
 & & \downarrow{\scriptstyle \xi'} & & \downarrow{\scriptstyle \xi} & & \downarrow{\scriptstyle \xi''} & & \\
0 & \longrightarrow & A & \xrightarrow{\chi'_{\omega_1}} & R \oplus \langle\mathbb{Z}\rangle & \xrightarrow{\psi'_{\omega_1}} & \langle\mathbb{Z}/p\rangle & \longrightarrow & 0.
\end{array}$$

An argument like that used for the $i > 1$ case implies that the map ξ vanishes on the summand $\langle\mathbb{Z}\rangle$ of its domain. The map ψ'_ω is, however, the projection onto this summand of Q_ω. Moreover, Proposition 1.10(k) implies that the map ξ'' is an epimorphism. Thus the composite along the top and right edge of the right square is a epimorphism when restricted to the summand $\langle\mathbb{Z}\rangle$. However, the composite along the left edge and bottom of this same square is zero when restricted to $\langle\mathbb{Z}\rangle$. This contradiction implies that $Q_\omega \neq \langle\mathbb{Z}\rangle \oplus L$ if $i = 1$. □

Proposition 4.12 and most of Proposition 4.14 follow easily from these results.

PROOFS OF PROPOSITIONS 4.12 AND 4.14. Observe that sequence (4.1) mentioned in these two propositions is just the middle row of diagram (5.4). Proposition 5.4 establishes the existence of an element ω'_{n+1} of $RO(G)$ with the properties claimed for it in Proposition 4.12. Given this element ω'_{n+1}, Proposition 5.6 can be applied to establish the assertion of Proposition 4.14 about sequence (4.1) in dimension ω_n. Once this claim is verified, Proposition 5.7 can be applied to establish the existence of an element ω'_n of $RO(G)$ with the properties claimed for it in Proposition 4.12. By continuing to apply Propositions 5.6 and 5.7 in an alternating fashion, we can establish the existence of the remaining ω'_i required by Proposition 4.12 and verify the claims of Proposition 4.14 about sequence (4.1). This completes the proof of Proposition 4.12. The only part of Proposition 4.14 which remains unproven is its claim about the map $\theta_{\omega_i} : R \oplus \langle\mathbb{Z}\rangle \longrightarrow R \oplus \langle\mathbb{Z}\rangle$. It is easy to check that there are no nonzero maps from R to $\langle\mathbb{Z}\rangle$ or from $\langle\mathbb{Z}\rangle$ to R. Thus, the map θ_{ω_i} must be of the form $f \oplus g$ for some maps $f : R \longrightarrow R$ and $g : \langle\mathbb{Z}\rangle \longrightarrow \langle\mathbb{Z}\rangle$. We need to prove that each of f and g is $\pm id$. Recall that the domain of θ is the free H_*-module $J' = \bigoplus_{1 \leq i \leq n+1} \Sigma^{\omega'_i} H_*$. It is easy to verify the claim of Proposition 4.14 that the R and $\langle\mathbb{Z}\rangle$ in the domain of θ_{ω_i} come from the generators of J' in the dimensions ω'_i and ω'_{i+1}, respectively. Thus, to prove the claim about θ_{ω_i}, it suffices to understand the restrictions of θ to the summands J'_i and J'_{i+1} coming from these two generators. Denote these restrictions by θ^i and θ^{i+1}, respectively. Recall that θ^i was defined by requiring that $\theta^i_{\omega'_i}$ be the identity map from $(J'_i)_{\omega'_i} = A$ to $Q_{\omega'_i} = A$.

To see that $f = \pm\, id$, consider the commuting square

$$\begin{array}{ccc} (J'_i)_{\omega'_i} & \xrightarrow{\hat{\xi}} & (J'_i)_{\omega_i} \\ {\scriptstyle \theta^i_{\omega'_i}}\downarrow\,{\scriptstyle =} & & \downarrow{\scriptstyle \theta^i_{\omega_i}} \\ Q_{\omega'_i} & \xrightarrow{\xi} & Q_{\omega_i} \end{array}$$

in which the horizontal maps come from multiplication by the generator $\xi_{\omega_i - \omega'_i}$ of $H_{\omega_i - \omega'_i}(G/G)$. Note that f is the composite of $\theta^i_{\omega_i}$ and the projection of Q_{ω_i} onto its R summand. The maps $\hat{\xi}$ and $\theta^i_{\omega'_i}$ take the generator μ' of $(J'_i)_{\omega'_i} = A$ to the generators ξ'_i of $(J'_i)_{\omega_i} = R$ and μ of $Q_{\omega'_i} = A$, respectively. Thus, to show that the map f is $\pm\, id$, it suffices to show that the map ξ in this square takes the generator μ of $Q_{\omega'_i} = A$ to $(\pm\,\xi_i, 0)$, where ξ_i is the generator of the R summand of $Q_{\omega_i} = R \oplus \langle\mathbb{Z}\rangle$. This map ξ is, essentially, the middle vertical map in one of the two main diagrams occurring the proof of Proposition 5.7. The appropriate one of the two depends on whether $i > 1$ or $i = 1$. In that proof we were arguing by contradiction and so assuming that $Q_{\omega'_i}$ was $\langle\mathbb{Z}\rangle \oplus L$ rather than A.

Redrawing the first of those two diagrams (the one for $i > 1$) with the correct value for $Q_{\omega'_i}$, we obtain the diagram

$$\begin{array}{ccccccccc} 0 & \longrightarrow & \langle\mathbb{Z}\rangle \oplus L & \xrightarrow{\chi'_{\omega'_i}} & A & \xrightarrow{\psi'_{\omega'_i}} & \langle\mathbb{Z}/p\rangle & \longrightarrow & 0 \\ & & {\scriptstyle \xi'}\downarrow & & \downarrow{\scriptstyle \xi} & & {\scriptstyle \cong}\downarrow{\scriptstyle \xi''} & & \\ 0 & \longrightarrow & A & \xrightarrow{\chi'_{\omega_i}} & R \oplus \langle\mathbb{Z}\rangle & \xrightarrow{\psi'_{\omega_i}} & \langle\mathbb{Z}/p\rangle & \longrightarrow & 0. \end{array}$$

At G/e, each of the four corners of the left square in this diagram is a copy of $\mathbb{Z}$, and the two horizontal maps in that square must be isomorphisms at G/e by exactness. It is easy to check that the left vertical map is also an isomorphism at G/e. From this it follows that $(\xi(G/G))(\mu) = (\pm\,\xi_i, x)$ for some $x \in \langle\mathbb{Z}\rangle(G/G) = \mathbb{Z}$. However, the map ξ must factor through R by an argument like that used to show the vanishing of ξ on one summand in the proof of Proposition 5.7. Since there are no nonzero maps from R to $\langle\mathbb{Z}\rangle$, it follows that the component of ξ going into the summand $\langle\mathbb{Z}\rangle$ of its range must be zero. Thus, $x = 0$. It follows that $f = \pm\, id$ if $i > 1$. The argument for the case $i = 1$ requires redrawing the other diagram in the proof of Proposition 5.7, but is essentially identical thereafter.

To see that $g = \pm\, id$, consider the commuting square

$$\begin{array}{ccc} (J'_{i+1})_{\omega'_{i+1}} & \xrightarrow{\hat{\epsilon}} & (J'_{i+1})_{\omega_i} \\ {\scriptstyle \theta^{i+1}_{\omega'_{i+1}}}\downarrow\,{\scriptstyle =} & & \downarrow{\scriptstyle \theta^{i+1}_{\omega_i}} \\ Q_{\omega'_{i+1}} & \xrightarrow{\epsilon} & Q_{\omega_i} \end{array}$$

in which the horizontal maps come from multiplication by the generator $\epsilon_{\omega_i - \omega'_{i+1}}$ of $H_{\omega_i - \omega'_{i+1}}(G/G)$. Note that g is the composite of $\theta^{i+1}_{\omega_i}$ and the projection of Q_{ω_i} onto its $\langle\mathbb{Z}\rangle$ summand. The maps $\hat{\epsilon}$ and $\theta^{i+1}_{\omega'_{i+1}}$ take the generator μ' of $(J'_{i+1})_{\omega'_{i+1}} = A$

to the generators ϵ' of $(J'_{i+1})_{\omega_i} = \langle \mathbb{Z} \rangle$ and μ of $Q_{\omega'_{i+1}} = A$, respectively. Thus, to show that the map g is $\pm\, id$, it suffices to show that the map ϵ in this square takes the generator μ of $Q_{\omega'_{i+1}} = A$ to $(0, \pm\epsilon_i)$, where ϵ_i is the generator of the $\langle \mathbb{Z} \rangle$ summand of $Q_{\omega_i} = R \oplus \langle \mathbb{Z} \rangle$. This map ϵ is the middle vertical map in one of the two main diagrams occurring the proof of Proposition 5.6. The appropriate one of the two depends on whether $i < n$ or $i = n$. As in the proof of Proposition 5.7, these two diagrams were drawn with an incorrect assumption about one of the entries in order to prove that incorrectness.

Correcting the first of these diagrams (which applies for $i < n$), we obtain the diagram

$$\begin{array}{ccccccccc}
0 & \longrightarrow & \langle \mathbb{Z} \rangle \oplus L & \xrightarrow{\chi'_{\omega'_{i+1}}} & A & \xrightarrow{\psi'_{\omega'_{i+1}}} & \langle \mathbb{Z}/p \rangle & \longrightarrow & 0 \\
 & & \downarrow{\scriptstyle \epsilon'} & & \downarrow{\scriptstyle \epsilon} & & \cong \downarrow{\scriptstyle \epsilon''} & & \\
0 & \longrightarrow & A & \xrightarrow{\chi'_{\omega_i}} & R \oplus \langle \mathbb{Z} \rangle & \xrightarrow{\psi'_{\omega_i}} & \langle \mathbb{Z}/p \rangle & \longrightarrow & 0.
\end{array}$$

Since the map ϵ vanishes at G/e, the component of this map going into the summand R of its range must be zero. Chasing the element $(1, 0)$ of $(\langle \mathbb{Z} \rangle \oplus L)(G/G)$ around this diagram in much the same way that it was chased around the analogous diagram in the proof of Proposition 5.6, one obtains fairly easily that $(\epsilon(G/G))(\mu) = (0, \pm\epsilon_i)$. Thus, $g = \pm\, id$ if $i < n$. For the case $i = n$, the other diagram in the proof of Proposition 5.6 must be redrawn correctly. Then chasing the generator 1 of $\langle \mathbb{Z} \rangle(G/G)$ around this diagram, much as in the proof of Proposition 5.6, gives that $(\epsilon(G/G))(\mu) = (0, \pm\epsilon_n)$. This completes the proof that $g = \pm\, id$ for all i. $\square$

CHAPTER 6

Dimension-shifting long exact sequences

In this chapter, we assume that the elements $\omega_1, \omega_2, \ldots, \omega_n$ of $RO(G)$ form a ramp of length n, $V \in RO(G)$ bounds this ramp, and the elements $\omega'_1, \omega'_2, \ldots, \omega'_{n+1}$ of $RO(G)$ form a V-shifted ramp as described in Definition 2.3. These elements of $RO(G)$ are also assumed to be even-dimensional and space-like. We assume further that J is a free H_*-module with n G/G-generators in the dimensions ω_i, J' is a free H_*-module with $n+1$ G/G-generators in the V-shifted dimensions ω'_j, and N is a free H_*-module having one G/G-generator in dimension V. Our primary goal here is to prove Proposition 4.7, which characterizes those maps $\bar{\chi} : J \longrightarrow J'$, $\bar{\psi} : J' \longrightarrow N$, and $\bar{\partial} : N \longrightarrow \Sigma J$ for which the sequence

$$\cdots \longrightarrow J \xrightarrow{\bar{\chi}} J' \xrightarrow{\bar{\psi}} N \xrightarrow{\bar{\partial}} \Sigma J \longrightarrow \cdots \tag{6.1}$$

is a long exact sequence. Throughout this chapter, we refer to any sequence of the form (6.1), regardless of whether it is exact, as a candidate sequence. An exact sequence of this form is referred to as a dimension-shifting long exact sequence.

Our proof of Proposition 4.7 is a three stage induction argument. The first stage of this argument, carried out in Section 6.2, is an induction on the number n of generators of J. This number is called the complexity of the sequence. The other two stages are on the two differences $|\omega_n^G - V^G|$ and $|V - \omega_n|$, which we refer to as the horizontal and vertical spreads of the sequence. These two stages are carried out in Section 6.4. Our induction arguments reduce the proof of the proposition to the special case of a candidate sequence with minimal complexity and minimal spread. This special case is handled in Section 6.3. In the final section of this chapter, we prove Proposition 4.9, which describes the only constraint on the dimensions ω_1, $\omega_2, \ldots, \omega_n, \omega'_1, \omega'_2, \ldots, \omega'_{n+1}$, and V which must be satisfied in order for there to be an associated dimension-shifting long exact sequence. Before going into these arguments, we describe a number of general properties of candidate sequences in Section 6.1.

6.1. Preliminary observations about dimension-shifting sequences

The restrictions imposed on the dimensions of the generators of N, J, and J' have a number of implications for the behavior of sequences of the form (6.1). These implications are explored in this section. In particular, we consider what the exactness of a sequence of this form tells us about the maps $\bar{\chi}$, $\bar{\psi}$, and $\bar{\partial}$. One goal of this discussion is to establish the necessity of the four conditions for exactness given in Proposition 4.7. A second goal is to describe some properties of sequences of complexity one ($n = 1$) which are used in the proof of the sufficiency of those four conditions. We begin with some observations about the vanishing of the composites in a candidate sequence.

LEMMA 6.1. *In any sequence of the form* (6.1),

(i) *the composite* $\bar{\psi} \circ \bar{\chi}$ *is zero if and only if the composites* $\bar{\psi}_{\omega_i} \circ \bar{\chi}_{\omega_i}$ *are zero for* $1 \leq i \leq n$.
(ii) *the composite* $\bar{\partial} \circ \bar{\psi}$ *is zero.*
(iii) *the composite* $\Sigma\bar{\chi} \circ \bar{\partial}$ *is zero if and only if the composite* $(\Sigma\bar{\chi})_V \circ \bar{\partial}_V$ *is zero.*
(iv) *the composite* $\Sigma\bar{\chi} \circ \bar{\partial}$ *is zero if the complexity* n *is 1.*

For the remainder of this section, we consider only exact sequences of the form (6.1), and investigate the implications of that exactness for the maps $\bar{\chi}$, $\bar{\psi}$, and $\bar{\partial}$.

PROPOSITION 6.2. *In an exact sequence of the form* (6.1), *the maps* $\bar{\chi} : J \longrightarrow J'$ *and* $\bar{\psi} : J' \longrightarrow N$ *are constructed from standard shift maps. Further, each component* $\bar{\partial}_i : N \longrightarrow \Sigma^{\omega_i+1} H_*$ *of the map* $\bar{\partial} : N \longrightarrow \Sigma J$ *is nonzero.*

PROOF. To prove this result, we must examine a sequence of the form (6.1) in the dimensions of the generators of J, J', and N. In the dimension ω_i of a generator of J, this sequence has the form

$$0 \longrightarrow A \xrightarrow{\bar{\chi}_{\omega_i}} R \oplus \langle \mathbb{Z} \rangle \xrightarrow{\bar{\psi}_{\omega_i}} \langle \mathbb{Z}/p \rangle \longrightarrow 0.$$

Here, the A comes from the generator of J in dimension ω_i, and R and $\langle \mathbb{Z} \rangle$ come from the generators of J' in dimensions ω_i' and ω_{i+1}', respectively. The remaining generators of J and J' contribute nothing in this dimension. Lemma 12.1 implies that the components $\bar{\chi}_{i,i}$ and $\bar{\chi}_{i,i+1}$ of the map $\bar{\chi}$ are standard shift maps. Since these components of $\bar{\chi}$ are the only ones that can be nonzero, it follows that $\bar{\chi}$ is constructed from standard shift maps.

In dimension ω_i', for $i \neq 1, n+1$, our sequence has the form

$$0 \longrightarrow \langle \mathbb{Z} \rangle \oplus L \xrightarrow{\bar{\chi}_{\omega_i'}} A \xrightarrow{\bar{\psi}_{\omega_i'}} \langle \mathbb{Z}/p \rangle \longrightarrow 0.$$

For $i = 1, n+1$, it has the forms

$$0 \longrightarrow L \xrightarrow{\bar{\chi}_{\omega_1'}} A \xrightarrow{\bar{\psi}_{\omega_1'}} \langle \mathbb{Z} \rangle \longrightarrow 0$$

and

$$0 \longrightarrow \langle \mathbb{Z} \rangle \xrightarrow{\bar{\chi}_{\omega_{n+1}'}} A \xrightarrow{\bar{\psi}_{\omega_{n+1}'}} R \longrightarrow 0$$

respectively. In these sequences, the $\langle \mathbb{Z} \rangle$ and L terms are contributed by the generators of J in dimensions ω_{i-1} and ω_i, respectively. The A terms are contributed by the generator of J' in dimension ω_i'. The remaining generators of J and J' contribute nothing in this dimension. From this description, it follows directly that all of the components of $\bar{\psi}$ which can be nonzero are surjective in the critical dimension and are therefore standard shift maps.

In the dimension V of the generator of N, our sequence has the form

$$\cdots \longrightarrow A \xrightarrow{\bar{\partial}_V} \bigoplus_{1 \leq i \leq n} \langle \mathbb{Z}/p \rangle \xrightarrow{\bar{\chi}_{V-1}} \bigoplus_{1 < j \leq n} \langle \mathbb{Z}/p \rangle \longrightarrow 0.$$

Clearly $\bar{\partial}_V$ must be nonzero if this sequence is exact. Thus, by Lemma 4.10, each component of $\bar{\partial}$ is nonzero. □

For the remainder of this section, we consider only sequences of complexity one. In this case, let $J_i' = \Sigma^{\omega_i'} H_*$, for $i = 1, 2$, so that $J' = J_1' \oplus J_2'$. Also, let $\bar{\chi}_i : J \longrightarrow J_i'$ and $\bar{\psi}_i : J_i' \longrightarrow N$ denote the components of the two maps $\bar{\chi}$ and $\bar{\psi}$.

PROPOSITION 6.3. *Assume that $n = 1$ and that the sequence*

$$\cdots \longrightarrow J \xrightarrow{\bar{\chi}} J' \xrightarrow{\bar{\psi}} N \xrightarrow{\bar{\partial}} \Sigma J \longrightarrow \cdots$$

is exact. Then any of the following changes to the maps $\bar{\chi}$, $\bar{\psi}$, and $\bar{\partial}$ yields another long exact sequence:

(i) *replacing $\bar{\partial}$ with any other nonzero map $N \longrightarrow \Sigma J$*
(ii) *replacing any two of the maps $\bar{\chi}_1$, $\bar{\chi}_2$, $\bar{\psi}_1$, and $\bar{\psi}_2$ with their negatives*
(iii) *if $p = 2$, replacing any one of the maps $\bar{\chi}_1$, $\bar{\chi}_2$, $\bar{\psi}_1$, and $\bar{\psi}_2$ with its negative.*

PROOF. For part (i), note that the map $\bar{\partial}$ must be nonzero by Proposition 6.2 and that the collection of maps from N to ΣJ is a cyclic group of order p by Lemma 1.12(a). Thus, any two nonzero maps from N to ΣJ are multiples of each other. Moreover, in any dimension ω where the map $\bar{\partial}$ is nonzero, its target is either $\langle \mathbb{Z}/p \rangle$ or L_- (which can occur only if $p = 2$). In either case, $\bar{\partial}_\omega(G/G)$ is surjective, and $\bar{\partial}_\omega(G/e)$ is zero. Replacing $\bar{\partial}$ by a nonzero multiple can therefore alter neither its image nor its kernel. For part (ii), observe that, if $\bar{\chi}_1$ and $\bar{\psi}_1$ are replaced by their negatives, then the new sequence can be compared to the old via the identity maps on J and N and the map $-1 \oplus 1 : J_1' \oplus J_2' \longrightarrow J_1' \oplus J_2'$. The exactness of the new sequence follows immediately from this comparison. Similarly, replacing $\bar{\chi}_2$ and $\bar{\psi}_2$ by their negatives also yields a new exact sequence. The new sequence obtained by replacing $\bar{\chi}$ and $\bar{\partial}$ by their negatives must be exact because we can compare it to the original one via the identity maps on J' and N and the map -1 on J. However, since $\bar{\partial}$ can be replaced by any nonzero map, it follows that replacing only $\bar{\chi}$ by its negative also produces a long exact sequence. Analogously, replacing $\bar{\psi}$ by its negative produces a long exact sequence. By combining pairs of these allowed sign changes, any other change of signs on exactly two of $\bar{\chi}_1$, $\bar{\chi}_2$, $\bar{\psi}_1$, and $\bar{\psi}_2$ can be accomplished. Thus, any change of exactly two signs does not alter the exactness of the sequence. Now assume that $p = 2$. For any $\alpha \in RO(G)$, at least one of the four Mackey functors J_α, $(J_1')_\alpha$, $(J_2')_\alpha$, and N_α is either $\langle \mathbb{Z}/2 \rangle$ or 0. It is impossible to tell whether the sign has been changed on a map into or out of either $\langle \mathbb{Z}/2 \rangle$ or 0. Thus, in any given dimension, a sequence obtained from our original sequence by changing the sign on exactly one of the maps $\bar{\chi}_1$, $\bar{\chi}_2$, $\bar{\psi}_1$, and $\bar{\psi}_2$ is indistinguishable from some sequence obtained by changing the signs on exactly two of these four maps. This indistinguishability implies the desired exactness. □

COROLLARY 6.4. *Assume $n = 1$, and let $\bar{\chi} : J \longrightarrow J'$, $\bar{\psi} : J' \longrightarrow N$, and $\bar{\partial} : N \longrightarrow \Sigma J$ be maps satisfying the four conditions in Proposition 4.7. Assume further that there are maps $\widehat{\chi} : J \longrightarrow J'$, $\widehat{\psi} : J' \longrightarrow N$, and $\widehat{\partial} : N \longrightarrow \Sigma J$ such that the sequence*

$$\cdots \longrightarrow J \xrightarrow{\widehat{\chi}} J' \xrightarrow{\widehat{\psi}} N \xrightarrow{\widehat{\partial}} \Sigma J \longrightarrow \cdots$$

is exact. Then the sequence

$$\cdots \longrightarrow J \xrightarrow{\bar{\chi}} J' \xrightarrow{\bar{\psi}} N \xrightarrow{\bar{\partial}} \Sigma J \longrightarrow \cdots$$

is also exact.

PROOF. Part (i) of Proposition 6.3 allow us to assume that $\bar{\partial} = \widehat{\partial}$. By assumption, the maps $\bar{\chi}_i : J \longrightarrow J'_i$ and $\bar{\psi}_i : J'_i \longrightarrow N$, for $i = 1, 2$, are standard shift maps. Proposition 6.2 indicates that the maps $\widehat{\chi}_i : J \longrightarrow J'_i$ and $\widehat{\psi}_i : J'_i \longrightarrow N$ must also be standard shift maps. Thus, $\bar{\chi}_i = \pm\widehat{\chi}_i$ and $\bar{\psi}_i = \pm\widehat{\psi}_i$. If $p = 2$, then we are done by part (iii) of Proposition 6.3. Thus, we may assume that p is odd. If an even number of the maps $\bar{\chi}_1$, $\bar{\chi}_2$, $\bar{\psi}_1$, and $\bar{\psi}_2$ are the negatives of the maps $\widehat{\chi}_1$, $\widehat{\chi}_2$, $\widehat{\psi}_1$, and $\widehat{\psi}_2$, then part (ii) of Proposition 6.3 implies the desired exactness. If an odd number of the four maps are negatives, then the composites

$$A \xrightarrow{\widehat{\chi}_{\omega_1}} R \oplus \langle \mathbb{Z} \rangle \xrightarrow{\widehat{\psi}_{\omega_1}} \langle \mathbb{Z}/p \rangle$$

and

$$A \xrightarrow{\bar{\chi}_{\omega_1}} R \oplus \langle \mathbb{Z} \rangle \xrightarrow{\bar{\psi}_{\omega_1}} \langle \mathbb{Z}/p \rangle$$

obtained by looking at our two long sequences in dimension ω_1 cannot both be zero. Since this contradicts our assumptions about the two sequences, an odd number of sign differences is not possible. □

The following special case of Proposition 4.9 can be coupled with the corollary above to simplify significantly the proof of Proposition 4.7 for sequences of complexity one.

LEMMA 6.5. *Assume that $n = 1$. There are maps $\bar{\chi} : J \longrightarrow J'$ and $\bar{\psi} : J' \longrightarrow N$, constructed from standard shift maps, such that the composite*

$$J_{\omega_1} \xrightarrow{\bar{\chi}_{\omega_1}} J'_{\omega_1} \xrightarrow{\bar{\psi}_{\omega_1}} N_{\omega_1}$$

is zero if and only if

$$d_{(V+\omega_1-\omega'_1-\omega'_2)} \equiv \pm 1 \mod p.$$

PROOF. Note that J_{ω_1}, $(J'_1)_{\omega'_1}$, and $(J'_2)_{\omega'_2}$ are all copies of the Burnside ring A, and let μ, μ'_1, and μ'_2 be the standard generators of these three Mackey functors. If the maps $\bar{\chi}$ and $\bar{\psi}$ are constructed from standard shift maps, then there are integers e_1, e_2, e'_1, and e'_2, each of which is ± 1, such that

$$(\bar{\chi}_1)_{\omega_1}(\mu) = e_1 \xi_{\omega_1-\omega'_1} \qquad (\bar{\chi}_2)_{\omega_1}(\mu) = e_2 \epsilon_{\omega_1-\omega'_2}$$
$$(\bar{\psi}_1)_{\omega'_1}(\mu'_1) = e'_1 \epsilon_{\omega'_1-V} \qquad (\bar{\chi}_2)_{\omega_1}(\mu'_2) = e'_2 \xi_{\omega'_2-V}.$$

Here $\xi_{\omega_1-\omega'_1}$, $\epsilon_{\omega_1-\omega'_2}$, $\epsilon_{\omega'_1-V}$, and $\xi_{\omega'_2-V}$ are the standard generators of $(J'_1)_{\omega_1}$, $(J'_2)_{\omega_1}$, $N_{\omega'_1}$, and $N_{\omega'_2}$ respectively. The map $\bar{\psi}_{\omega_1} \circ \bar{\chi}_{\omega_1}$ is zero if and only if

$$(\bar{\psi}_{\omega_1} \circ \bar{\chi}_{\omega_1})(\mu) = 0.$$

However,

$$\begin{aligned}(\bar{\psi}_{\omega_1} \circ \bar{\chi}_{\omega_1})(\mu) &= e_1 e'_1 \epsilon_{\omega'_1-V} \xi_{\omega_1-\omega'_1} + e_2 e'_2 \epsilon_{\omega_1-\omega'_2} \xi_{\omega'_2-V} \\ &= (e_1 e'_1 + e_2 e'_2 d_{(V+\omega_1-\omega'_1-\omega'_2)}) \epsilon_{\omega'_1-V} \xi_{\omega_1-\omega'_1}.\end{aligned}$$

Here, the second equality follows from Proposition 1.10(l). Since each of e_1, e_2, e'_1, and e'_2 is ± 1, $(\bar{\psi}_{\omega_1} \circ \bar{\chi}_{\omega_1})(\mu)$ can be zero if and only if $d_{(V+\omega_1-\omega'_1-\omega'_2)} \equiv \pm 1 \mod p$. □

Together this lemma and the corollary preceding it reduce the sufficiency part of the proof of Proposition 4.7 for a sequence of complexity one to showing that, whenever ω_1, ω_1', ω_2', and V are even-dimensional space-like elements of $RO(G)$ for which $d_{(V+\omega_1-\omega_1'-\omega_2')} \equiv \pm 1 \mod p$, there is at least one choice for the maps $\bar{\chi} : J \longrightarrow J'$, $\bar{\psi} : J' \longrightarrow N$, and $\bar{\partial} : N \longrightarrow \Sigma J$ which makes sequence (6.1) exact.

6.2. The reduction to complexity one dimension-shifting sequences

Here, we give an inductive argument which reduces the proof of Proposition 4.7 to the case of sequences of complexity one. Thus, throughout this section, we assume that the complexity n of our sequence is at least 2 and that the maps $\bar{\chi} : J \longrightarrow J'$, $\bar{\psi} : J' \longrightarrow N$, and $\bar{\partial} : N \longrightarrow \Sigma J$ satisfy the four conditions in the proposition. Our goal is to show that sequence (6.1) is exact. We do this by comparing this sequence to two other sequences of complexities 1 and $n-1$, respectively. These other two sequences satisfy the conditions of the proposition, and so we may assume that they are exact.

For $1 \le i \le n$ and $1 \le j \le n+1$, let $J_i = \Sigma^{\omega_i} H_*$ and $J_j' = \Sigma^{\omega_j'} H_*$, so that $J = \oplus_i J_i$ and $J' = \oplus_j J_j'$. Let $\chi'' : J_n \longrightarrow J_n' \oplus J_{n+1}'$ be the map obtained from $\bar{\chi}$ by restriction to the summand J_n of its domain and projection onto the summand $J_n' \oplus J_{n+1}'$ of its range. The map χ'' is constructed from standard shift maps since the maps $\bar{\chi}$ and $\bar{\psi}$ are assumed to satisfy condition (i) of Proposition 4.7. Let $\bar{\psi}_j : J_j' \longrightarrow N$ be the jth component of $\bar{\psi}$. Condition (i) also implies that the map $\bar{\psi}_n$ is nonzero. This map is determined by its value in dimension ω_n', which has the form $(\bar{\psi}_n)_{\omega_n'} : A \longrightarrow \langle \mathbb{Z}/p \rangle$. From parts (k) and (l) of Proposition 1.10, it follows that there are elements $\alpha, \beta \in RO(G)$ such that $(\bar{\psi}_n)_{\omega_n'}(\mu) = \pm \epsilon_\alpha \xi_\beta$. These two elements of $RO(G)$ satisfy the conditions

$$|\alpha| = |\omega_n'| - |V| \quad \text{and} \quad |\alpha^G| = 0$$
$$|\beta| = 0 \quad \text{and} \quad |\beta^G| = |(\omega_n')^G| - |V^G|.$$

Let $\bar{N} = \Sigma^{V+\beta} H_*$. Then $\bar{\psi}_n$ can be written as the composite

$$J_n' \xrightarrow{\psi_1''} \bar{N} \xrightarrow{\widehat{\psi}_n} N$$

in which the first map is multiplication by ϵ_α and the second is multiplication by $\pm \xi_\beta$. Note that both of these maps are standard shift maps. It follows from Proposition 1.10(g) that the map $\bar{\psi}_{n+1} : J_{n+1}' \longrightarrow N$ can be written as a composite of the form

$$J_{n+1}' \xrightarrow{\psi_2''} \bar{N} \xrightarrow{\widehat{\psi}_n} N$$

in which the first map is multiplication by $\pm \xi_{V+\beta-\omega_{n+1}'}$, and so is a standard shift map.

Together, the maps ψ_1'' and ψ_2'' give a map $\psi'' : J_n' \oplus J_{n+1}' \longrightarrow \bar{N}$ which is constructed from standard shift maps. For dimensional reasons, the composite

$$\bar{N} \xrightarrow{\widehat{\psi}_n} N \xrightarrow{\bar{\partial}} \Sigma J$$

factors through the inclusion $\Sigma i_n : \Sigma J_n \longrightarrow \Sigma J$ via a map $\partial'' : \bar{N} \longrightarrow \Sigma J_n$. Since $\bar{\partial}_n : N \longrightarrow \Sigma J_n$ is nonzero by condition (ii) of Proposition 4.7, Proposition 1.10(s) implies that the map ∂'' is also nonzero. The maps χ'', ψ'', and ∂'' fit into the commuting diagram

$$\begin{array}{ccccccccccc}
\cdots & \longrightarrow & J_n & \xrightarrow{\chi''} & J'_n \oplus J'_{n+1} & \xrightarrow{\psi''} & \bar{N} & \xrightarrow{\partial''} & \Sigma J_n & \longrightarrow & \cdots \\
 & & \downarrow{\scriptstyle i_n} & & \downarrow{\scriptstyle i} & & \downarrow{\scriptstyle \widehat{\psi}_n} & & \downarrow{\scriptstyle \Sigma i_n} & & \\
\cdots & \longrightarrow & J & \xrightarrow[\bar{\chi}]{} & J' & \xrightarrow[\bar{\psi}]{} & N & \xrightarrow[\bar{\partial}]{} & \Sigma J & \longrightarrow & \cdots .
\end{array} \tag{6.2}$$

We have already observed that the top row of this diagram satisfies conditions (i) and (ii) of Proposition 4.7. Part (iv) of Lemma 6.1 indicates that it satisfies condition (iv). By looking at parts (g) and (k) of Proposition 1.10, we can argue that the map $\widehat{\psi}_n$ is an isomorphism in dimension ω_n. From this, it follows that the top row also satisfies condition (iii) of the proposition. Thus, by Proposition 6.10 in Section 6.4, the top row is a long exact sequence.

Let $\bar{J} = \oplus_{i=1}^{n-1} J_i$ and $\bar{J}' = \oplus_{j=1}^{n-1} J'_j$. By adding $\bar{J}'$ to the $J'_n \oplus J'_{n+1}$ and $\bar{N}$ terms of the top row of diagram (6.2), we obtain the long exact sequence

$$\cdots \longrightarrow J_n \xrightarrow{\bar{\chi}_n} J' \xrightarrow{1\oplus\psi''} \bar{J}' \oplus \bar{N} \xrightarrow{(0,\partial'')} \Sigma J_n \longrightarrow \cdots$$

in which $\bar{\chi}_n$ is just the restriction of $\bar{\chi}$ to J_n.

Let

$$\widehat{\chi} : \bar{J} \longrightarrow \bar{J}' \oplus \bar{N}$$

be the composite of the map $1 \oplus \psi'' : J' \longrightarrow \bar{J}' \oplus \bar{N}$ and the restriction of the map $\bar{\chi} : J \longrightarrow J'$ to the summand $\bar{J}$ of J. Also, let

$$\widehat{\psi} : \bar{J}' \oplus \bar{N} \longrightarrow N$$

be the map formed from the restriction of $\bar{\psi} : J' \longrightarrow N$ to the summand $\bar{J}'$ of J' and the map $\widehat{\psi}_n : \bar{N} \longrightarrow N$. Further, let

$$\widehat{\partial} : N \longrightarrow \Sigma\bar{J}$$

be the composite of the map $\bar{\partial} : N \longrightarrow \Sigma J$ and the projection $\Sigma\pi : \Sigma J \longrightarrow \Sigma\bar{J}$ onto the summand $\Sigma\bar{J}$ of ΣJ. Then we have the commuting diagram

$$\begin{array}{ccccccccccc}
 & & & & & & \vdots & & & & \\
 & & & & & & \downarrow & & & & \\
 & & & & & \overset{\bar{\chi}|_{\bar{J}}}{\swarrow} & \bar{J} & & & & \\
 & & & & & & \downarrow{\scriptstyle \widehat{\chi}} & & & & \\
\cdots & \longrightarrow & J_n & \xrightarrow[\bar{\chi}_n]{} & J' & \xrightarrow[1\oplus\psi'']{} & \bar{J}' \oplus \bar{N} & \xrightarrow[(0,\partial'')]{} & \Sigma J_n & \longrightarrow & \cdots \\
 & & \downarrow{\scriptstyle i_n} & & \downarrow{\scriptstyle =} & & \downarrow{\scriptstyle \widehat{\psi}} & & \downarrow{\scriptstyle \Sigma i_n} & & \\
\cdots & \longrightarrow & J & \xrightarrow[\bar{\chi}]{} & J' & \xrightarrow[\bar{\psi}]{} & N & \xrightarrow[\bar{\partial}]{} & \Sigma J & \longrightarrow & \cdots \\
 & & & & & & \downarrow{\scriptstyle \widehat{\partial}} & & \downarrow{\scriptstyle \pi} & & \\
 & & & & & & \Sigma\bar{J} & \xrightarrow[=]{} & \Sigma\bar{J} & & \\
 & & & & & & \downarrow & & & & \\
 & & & & & & \vdots & & & &
\end{array} \tag{6.3}$$

The vertical column in the center of this diagram is exactly the sort of sequence to which Proposition 4.7 applies. Moreover, since it is a sequence of complexity $n-1$, we may assume inductively that the proposition is valid. It follows easily from the definitions of the maps in the vertical column that conditions (i) and (ii) of the proposition are satisfied by the column. The composite

$$\bar{J} \xrightarrow{\widehat{\chi}} \bar{J}' \oplus \bar{N} \xrightarrow{\widehat{\psi}} N$$

is just the restriction of the composite

$$J \xrightarrow{\bar{\chi}} J' \xrightarrow{\bar{\psi}} N$$

to the summand $\bar{J}$ of J. Our assumption that the composite $\bar{\psi} \circ \bar{\chi}$ vanishes in dimension ω_i, for $1 \leq i \leq n$, therefore implies that $\widehat{\psi} \circ \widehat{\chi}$ vanishes in dimension ω_i for $1 \leq i \leq n-1$. Thus, condition (iii) of the proposition is satisfied. To see that the vertical column satisfies (iv) of the proposition, consider the diagram

$$\begin{array}{ccccc} N & \xrightarrow{\bar{\partial}} & \Sigma J & \xrightarrow{\Sigma\bar{\chi}} & \Sigma J' \\ & \searrow^{\widehat{\partial}} & & & \downarrow{\scriptstyle \Sigma(1 \oplus \psi'')} \\ & & \Sigma\bar{J} & \xrightarrow{\Sigma\widehat{\chi}} & \Sigma(\bar{J}' \oplus \bar{N}). \end{array}$$

The fact that the summand ΣJ_n is present in ΣJ but missing from $\Sigma\bar{J}$ might suggest that this diagram does not commute. However, for dimensional reasons, the portion $\bar{\partial}_n : N \longrightarrow \Sigma J_n$ of the map $\bar{\partial} : N \longrightarrow \Sigma J$ contributes nothing to the composite along the top and right-hand side of this diagram. It follows easily that the diagram does commute. By assumption, the composite $\Sigma\bar{\chi} \circ \bar{\partial}$ vanishes in dimension V. Thus, $\Sigma\widehat{\chi} \circ \widehat{\partial}$ also vanishes in dimension V, and condition (iv) of the proposition is satisfied.

We now know that the top row and the central vertical column of diagram (6.3) are exact, and must prove that the bottom row is exact. We have assumed that the bottom row satisfies conditions (iii) and (iv) of Proposition 4.7. Lemma 6.1 indicates that this implies the vanishing of the composites $\bar{\psi} \circ \bar{\chi}$ and $\Sigma\bar{\chi} \circ \bar{\partial}$. That lemma also asserts that the composite $\bar{\partial} \circ \bar{\psi}$ is zero. Thus, to complete the proof that the bottom row of diagram (6.3) is exact, it suffices to show that the kernel of each map is contained in the image of the previous map. We do this by chasing elements around the diagram. Even though this diagram is a diagram of Mackey functors, we may treat it as a diagram of abelian groups for the purpose of chasing elements — basically because the diagram is exact if and only if it is exact when evaluated at G/G and G/e. Exactness at N follows from an utterly routine diagram chase.

The key to establishing exactness at J and J' is the observation that $J = \bar{J} \oplus J_n$ and that, under this identification, the map $\bar{\chi} : J \longrightarrow J'$ is just the sum of the maps $\bar{\chi}|_{\bar{J}} : \bar{J} \longrightarrow J'$ and $\bar{\chi}_n : J_n \longrightarrow J'$. It is easy to see that, if $y \in J'$ such that $\bar{\psi}(y) = 0$, then there are elements $\bar{x} \in \bar{J}$ and $x_n \in J_n$ such that $y = (\bar{\chi}|_{\bar{J}})(\bar{x}) + \bar{\chi}_n(x_n)$. It follows that, if we regard the pair $(\bar{x}, x_n)$ as an element of J, then $\bar{\chi}(\bar{x}, x_n) = y$. Similarly, given $x \in J$ such that $\bar{\chi}(x) = 0$, regard x as a pair $(\bar{x}, x_n)$ with $\bar{x} \in \bar{J}$ and $x_n \in J_n$. The assertion that $\bar{\chi}(x) = 0$ is equivalent to the statement that $(\bar{\chi}|_{\bar{J}})(\bar{x}) = -\bar{\chi}_n(x_n)$. The exactness of the top row gives that $((1 \oplus \psi'') \circ \bar{\chi}_n)(x_n) = 0$. Thus, $\widehat{\chi}(\bar{x}) = 0$. The exactness of the vertical column then gives an element z of $\Sigma^{-1}N$ such that $(\Sigma^{-1}\widehat{\partial})(z) = \bar{x}$. It is easy to see that there is an element x_n' of J_n such that $x - (\Sigma^{-1}\bar{\partial})(z) = i_n(x_n')$. Since the composite $\bar{\chi} \circ \Sigma^{-1}\bar{\partial}$ is zero,

$\bar{\chi}(x - (\Sigma^{-1}\bar{\partial})(z)) = 0$, and so $\bar{\chi}_n(x'_n) = 0$. The exactness of the top row then gives an element w of $\Sigma^{-1}(\bar{J}' \oplus \bar{N})$ such that $(\Sigma^{-1}(0, \partial''))(w) = x'_n$. It follows that $x = (\Sigma^{-1}\bar{\partial})(z + (\Sigma^{-1}\widehat{\psi})(w))$. This completes our reduction of the proof of Proposition 4.7 to the case in which the complexity n is 1.

6.3. Sequences with minimal complexity and spread

The induction arguments presented in the previous section and the next section reduce the proof of the sufficiency part of Proposition 4.7 down to proving the following proposition and corollary.

PROPOSITION 6.6. *Assume that $n = 1$ and that $|\omega_1^G - V^G| = |V - \omega_1| = 2$. If $d_{V+\omega_1-\omega_1'-\omega_2'} \equiv \pm 1 \mod p$, then there exist maps $\widehat{\chi} : J \longrightarrow J'$, $\widehat{\psi} : J' \longrightarrow N$, and $\widehat{\partial} : N \longrightarrow \Sigma J$ such that the sequence*

$$\cdots \longrightarrow J \xrightarrow{\widehat{\chi}} J' \xrightarrow{\widehat{\psi}} N \xrightarrow{\widehat{\partial}} \Sigma J \longrightarrow \cdots$$

is a long exact sequence.

COROLLARY 6.7. *Assume that $n = 1$ and that $|\omega_1^G - V^G| = |V - \omega_1| = 2$. If*

(i) *$\bar{\chi}$ and $\bar{\psi}$ are constructed from standard shift maps,*
(ii) *the map $\bar{\partial} : N \longrightarrow \Sigma J$ is nonzero, and*
(iii) *the composite*

$$J_{\omega_1} \xrightarrow{\bar{\chi}_{\omega_1}} J'_{\omega_1} \xrightarrow{\bar{\psi}_{\omega_1}} N_{\omega_1}$$

is zero,

then the sequence

$$\cdots \longrightarrow J \xrightarrow{\bar{\chi}} J' \xrightarrow{\bar{\psi}} N \xrightarrow{\bar{\partial}} \Sigma J \longrightarrow \cdots$$

is a long exact sequence.

PROOF. We have assumed that the maps $\bar{\chi}$, $\bar{\psi}$, and $\bar{\partial}$ satisfy the first three conditions of Proposition 4.7. Part (iv) of Lemma 6.1 indicates that the fourth condition in that proposition is also satisfied. By Lemma 6.5, $d_{V+\omega_1-\omega_1'-\omega_2'} \equiv \pm 1 \mod p$. Thus, by Proposition 6.6, there are maps $\widehat{\chi} : J \longrightarrow J'$, $\widehat{\psi} : J' \longrightarrow N$, and $\widehat{\partial} : N \longrightarrow \Sigma J$ such that the sequence

$$\cdots \longrightarrow J \xrightarrow{\widehat{\chi}} J' \xrightarrow{\widehat{\psi}} N \xrightarrow{\widehat{\partial}} \Sigma J \longrightarrow \cdots$$

is exact. Corollary 6.4 now implies the asserted exactness. □

There are two possible approaches to proving Proposition 6.6. The most direct approach, which was taken in [**6**], is to select the maps $\widehat{\chi}$, $\widehat{\psi}$, and $\widehat{\partial}$ appropriately, and then prove the exactness of the sequence simply by examining it in all possible dimensions. However, this approach is quite tedious and requires intimate familiarity with both the additive and multiplicative structure of H_*. A shorter proof can be obtained by applying the main freeness result from [**12**] to appropriately selected stunted complex projective spaces. Certain cell-attaching long exact sequences for these spaces are sequences of exactly the desired form. The remainder of this section is devoted to proving the proposition via this second approach.

The first step in this approach is to note that we can reduce the proof of the proposition to a special case in which we have replaced the relatively unrestricted quadruple of elements V, ω_1, ω_1', and ω_2' of $RO(G)$ by a much more carefully selected quadruple. In particular, by desuspending the desired sequence by ω_1', we can reduce the proof of the proposition to the special case in which the quadruple V, ω_1, ω_1', and ω_2' has been replaced by the quadruple $V-\omega_1'$, $\omega_1-\omega_1'$, 0, and $\omega_2'-\omega_1'$. By Lemma 1.5, we can select nontrivial irreducible complex G-representations η and λ such that

$$d_{\eta-(V-\omega_1')} \equiv d_{2-\lambda-(\omega_1-\omega_1')} \equiv \pm 1 \mod p.$$

Lemma 1.14 then provides isomorphisms

$$\Sigma^{V-\omega_1'} H_* \cong \Sigma^{\eta} H_* \quad \text{and} \quad \Sigma^{\omega_1-\omega_1'} H_* \cong \Sigma^{2-\lambda} H_*$$

of H_*-modules which allow us to replace $V-\omega_1'$ and $\omega_1-\omega_1'$ by η and $2-\lambda$, respectively. The congruences determining η and λ can be coupled with the congruence $d_{\eta^{-1}-\eta} \equiv \pm 1 \mod p$ of Lemma 1.5 and the congruence $d_{V+\omega_1-\omega_1'-\omega_2'} \equiv \pm 1 \mod p$ assumed in the proposition to obtain the congruence

$$d_{2-\lambda+\eta^{-1}-(\omega_2'-\omega_1')} \equiv \pm 1 \mod p.$$

Applying Lemma 1.14 again gives us the isomorphism

$$\Sigma^{\omega_2'-\omega_1'} H_* \cong \Sigma^{2-\lambda+\eta^{-1}} H_*$$

of H_*-modules which allows us to replace $\omega_2'-\omega_1'$ by $2-\lambda+\eta^{-1}$. Thus, it suffices to prove the special case of Proposition 6.6 in which the quadruple of elements of $RO(G)$ is η, $2-\lambda$, 0, and $2-\lambda+\eta^{-1}$. Via suspension by λ, this special case is equivalent to the special case of the quadruple $\eta+\lambda$, 2, λ, and $2+\eta^{-1}$.

In the special case where $\eta=\lambda$, applying the results of [**12**] to a copy of $\mathbb{C}P^2$ with a linear action provides the desired long exact sequence.

LEMMA 6.8. *Let η be a nontrivial irreducible complex G-representation. Then, there are maps*

$$\widehat{\chi} : \Sigma^2 H_* \longrightarrow \Sigma^{\eta} H_* \oplus \Sigma^{2+\eta^{-1}} H_*,$$
$$\widehat{\psi} : \Sigma^{\eta} H_* \oplus \Sigma^{2+\eta^{-1}} H_* \longrightarrow \Sigma^{2\eta} H_*,$$

and

$$\widehat{\partial} : \Sigma^{2\eta} H_* \longrightarrow \Sigma^3 H_*$$

such that the sequence

$$\cdots \longrightarrow \Sigma^2 H_* \xrightarrow{\widehat{\chi}} \Sigma^{\eta} H_* \oplus \Sigma^{2+\eta^{-1}} H_* \xrightarrow{\widehat{\psi}} \Sigma^{2\eta} H_* \xrightarrow{\widehat{\partial}} \Sigma^3 H_* \longrightarrow \cdots$$

is exact.

PROOF. For any complex G-representation W, denote the associated complex projective space with a linear G-action by $P(W)$. Also, denote the trivial complex G-representation with complex dimension n by $n_{\mathbb{C}}$. By Proposition 3.1 of [**12**], the reduced homology of $P(2_{\mathbb{C}}+\eta^{-1})$ is a free H_*-module with generators in dimensions η and $2+\eta^{-1}$. This description of $\widetilde{H}_*^G(P(2_{\mathbb{C}}+\eta^{-1}); A)$ is obtained by viewing this space as being obtained from $P(1_{\mathbb{C}}+\eta^{-1})$ by attaching the obvious 4-cell. If, instead, we view this space as being obtained from $P(2_{\mathbb{C}})$ by attaching a different 4-cell, then the associated cell-attaching long exact sequence has the form

$$\cdots \longrightarrow \Sigma^2 H_* \longrightarrow \widetilde{H}_*^G(P(2_{\mathbb{C}}+\eta^{-1});A) \longrightarrow \Sigma^{2\eta} H_* \xrightarrow{\partial} \Sigma^3 H_* \longrightarrow \cdots .$$

Replacing $\widetilde{H}_*^G(P(2_{\mathbb{C}}+\eta^{-1});A)$ by the isomorphic H_*-module $\Sigma^{\eta} H_* \oplus \Sigma^{2+\eta^{-1}} H_*$ gives the desired long exact sequence. Note that the boundary map ∂ in this long exact sequence has to be nonzero since the free H_*-modules $\Sigma^2 H_* \oplus \Sigma^{2\eta} H_*$ and $\Sigma^{\eta} H_* \oplus \Sigma^{2+\eta^{-1}} H_*$ are obviously not isomorphic. □

If $\eta \neq \lambda$, then we must use a stunted projective space with a linear G-action to produce the desired long exact sequence. Observe that, since η and λ are both nontrivial irreducible complex G-representations, there is an integer k such that $\lambda = \eta^k$ and $1 < k < p$. Let W be the complex G-representation $1_{\mathbb{C}} + \eta + \ldots + \eta^{k-1}$. Note that neither λ nor $\eta^{-1} = \eta^{p-1}$ is contained in W. Applying the results of [**12**] to the stunted projective space $P(W + \eta^{-1} + 1_{\mathbb{C}})/P(W)$ provides the desired long exact sequence.

LEMMA 6.9. *Let η be a nontrivial irreducible complex G-representation, and k be an integer such that $1 < k < p$. Then, there are maps*

$$\widehat{\chi} : \Sigma^2 H_* \longrightarrow \Sigma^{\eta^k} H_* \oplus \Sigma^{2+\eta^{-1}} H_*,$$
$$\widehat{\psi} : \Sigma^{\eta^k} H_* \oplus \Sigma^{2+\eta^{-1}} H_* \longrightarrow \Sigma^{\eta+\eta^k} H_*,$$

and

$$\widehat{\partial} : \Sigma^{\eta+\eta^k} H_* \longrightarrow \Sigma^3 H_*$$

such that the sequence

$$\cdots \longrightarrow \Sigma^2 H_* \xrightarrow{\widehat{\chi}} \Sigma^{\eta^k} H_* \oplus \Sigma^{2+\eta^{-1}} H_* \xrightarrow{\widehat{\psi}} \Sigma^{\eta+\eta^k} H_* \xrightarrow{\widehat{\partial}} \Sigma^3 H_* \longrightarrow \cdots$$

is exact.

PROOF. Let $X = P(W + \eta^{-1} + 1_{\mathbb{C}})/P(W)$ and $\omega = \eta + \eta^2 + \ldots + \eta^{k-1}$. By Proposition 3.1 of [**12**], the reduced homology of X is a free H_*-module with generators in dimensions $\omega + \eta^k$ and $2 + \omega + \eta^{-1}$. This description of $\widetilde{H}_*^G(X;A)$ is obtained by viewing this space as being obtained from $P(W+\eta^{-1})/P(W)$ by attaching the appropriate $(2k+2)$-cell. If, instead, we view this space as being obtained from $P(W + 1_{\mathbb{C}})/P(W)$ by attaching a different $(2k+2)$-cell, then the associated cell-attaching long exact sequence has the form

$$\cdots \longrightarrow \Sigma^{2+\omega} H_* \longrightarrow \widetilde{H}_*^G(X;A) \longrightarrow \Sigma^{\omega+\eta^k+\eta} H_* \xrightarrow{\partial} \Sigma^{3+\omega} H_* \longrightarrow \cdots .$$

Replacing $\widetilde{H}_*^G(X;A)$ by the isomorphic H_*-module $\Sigma^{\omega+\eta^k} H_* \oplus \Sigma^{2+\omega+\eta^{-1}} H_*$ and desuspending by ω gives the desired long exact sequence. Note that the map ∂ in this sequence has to be nonzero since the free H_*-modules $\Sigma^{2+\omega} H_* \oplus \Sigma^{\omega+\eta^k+\eta} H_*$ and $\Sigma^{\omega+\eta^k} H_* \oplus \Sigma^{2+\omega+\eta^{-1}} H_*$ are obviously not isomorphic. □

Together Lemmas 6.8 and 6.9 provide a long exact sequence for every quadruple of the form $\eta + \lambda$, 2, λ, and $2 + \eta^{-1}$. Suspending by the appropriate element of $RO(G)$ and applying the appropriate isomorphisms from Lemma 1.14 then provides a long exact sequence for each quadruple V, ω_1, ω_1', and ω_2' satisfying the hypotheses of Proposition 6.6.

6.4. The reduction to sequences of minimal spread

Throughout this section, we assume that the complexity n of our sequences is one. Our goal in this section is to prove the following special case of Proposition 4.7.

PROPOSITION 6.10. *Assume $n = 1$. The sequence*

$$\cdots \longrightarrow J \xrightarrow{\bar{\chi}} J' \xrightarrow{\bar{\psi}} N \xrightarrow{\bar{\partial}} \Sigma J \longrightarrow \cdots \tag{6.4}$$

is a long exact sequence if and only if the following three conditions are satisfied:

(i) *$\bar{\chi}$ and $\bar{\psi}$ are constructed from standard shift maps*
(ii) *the map $\bar{\partial} : N \longrightarrow \Sigma J$ is nonzero*
(iii) *the composite*

$$J_{\omega_1} \xrightarrow{\bar{\chi}_{\omega_1}} J'_{\omega_1} \xrightarrow{\bar{\psi}_{\omega_1}} N_{\omega_1}$$

is zero.

The fourth condition which one might expect to see here is unnecessary by part (iv) of Lemma 6.1. Condition (iii) in this proposition is obviously necessary for exactness. Proposition 6.2 implies that the first two conditions are also necessary for exactness. The remainder of this section is devoted to proving that these three conditions are also sufficient. Thus, assume that $\bar{\chi} : J \longrightarrow J'$, $\bar{\psi} : J' \longrightarrow N$, and $\bar{\partial} : N \longrightarrow \Sigma J$ are maps satisfying the three conditions in the proposition. Note that, by Lemma 6.5,

$$d_{(V+\omega_1-\omega_1'-\omega_2')} \equiv \pm 1 \mod p.$$

Our proof is a two stage induction argument based on Corollary 6.7. In the first stage, we retain the restriction from the corollary that $|V - \omega_1| = 2$, but eliminate the constraint on $|\omega_1^G - V^G|$ by an induction on the size of $|\omega_1^G - V^G|$. Corollary 6.7 serves as both the base case of this induction and a key tool in proving the inductive step. In the second stage of the induction, we use the result from the first stage to eliminate the constraint on $|V - \omega_1|$ by an induction on the size of $|V - \omega_1|$.

For the first stage of our induction, we work with a quadruple V, ω_1, ω_1', and ω_2' of even-dimensional space-like elements $RO(G)$ such that $|V - \omega_1| = 2$. In order to apply our induction hypothesis, we wish to replace this quadruple of elements with two other quadruples having smaller horizonal spreads. Each of these new quadruples are formed by replacing a pair of the elements from the original quadruple by the elements $\omega_1' + 2$ and $2\omega_1' - V + 2$ of $RO(G)$. Figure 6.1 illustrates the relative positions of these six elements of $RO(G)$.

Let $\tilde{N} = \Sigma^{\omega_1'+2} H_*$ and $\tilde{J} = \Sigma^{2\omega_1' - V + 2} H_*$. Observe that the quadruple V, $2\omega_1' - V + 2$, ω_1', and $\omega_1' + 2$ (with $2\omega_1' - V + 2$ and ω_1' $+2$ taken as replacements for ω_1 and ω_2', respectively) satisfies the the hypotheses of Proposition 6.6. Thus, for appropriately chosen maps χ', ψ', and ∂', we have a dimension-shifting long exact sequence

$$\cdots \longrightarrow \tilde{J} \xrightarrow{\chi'} J_1' \oplus \tilde{N} \xrightarrow{\psi'} N \xrightarrow{\partial'} \Sigma \tilde{J} \longrightarrow \cdots . \tag{6.5}$$

Consider also the quadruple $\omega_1' + 2$, ω_1, $2\omega_1' - V + 2$, and ω_2' (with $\omega_1' + 2$ taken as a replacement for V and $2\omega_1' - V + 2$ taken as a replacement for ω_1'). Since

$$(\omega_1' + 2) + \omega_1 - (2\omega_1' - V + 2) - \omega_2' = V + \omega_1 - \omega_1' - \omega_2',$$

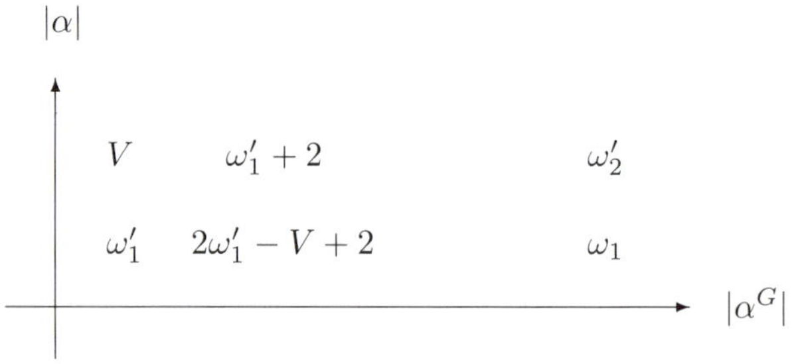

FIGURE 6.1. The six elements of $RO(G)$ used in stage one of the induction

Lemma 6.5 provides us with maps $\chi'' : J \longrightarrow \tilde{J} \oplus J_2'$ and $\psi'' : \tilde{J} \oplus J_2' \longrightarrow \tilde{N}$, constructed from standard shift maps, such that the composite $\psi''_{\omega_1} \circ \chi''_{\omega_1}$ is zero. Moreover, because

$$|\omega_1^G - (\omega_1' + 2)^G| < |\omega_1^G - V^G|,$$

our induction hypothesis allows us to assume that the sequence

$$\cdots \longrightarrow J \xrightarrow{\chi''} \tilde{J} \oplus J_2' \xrightarrow{\psi''} \tilde{N} \xrightarrow{\partial''} \Sigma J \longrightarrow \cdots \tag{6.6}$$

is exact provided the map ∂'' is nonzero.

Proposition 6.3 implies that we have a certain amount of flexibility in the choice of the maps χ', ψ', ∂', χ'', ψ'', and ∂'' in these two long exact sequences. We want to use that flexibility to select those maps in such a way that we can derive the exactness of sequence (6.4) from the exactness of these two sequences. Each of the maps $\bar{\chi}$, $\bar{\psi}$, χ', ψ', χ'', and ψ'' has two components, which we denote using subscripts (as in $\bar{\chi}_1$ and $\bar{\chi}_2$). These six components fit into the diagram

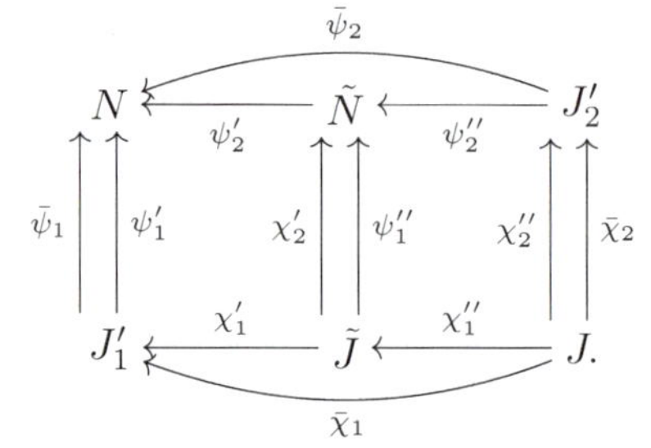

Since all of the maps in this diagram are standard shift maps, the maps in each parallel pair are either equal or negatives of each other. Moreover, the composites $\chi_1' \circ \chi_1''$ and $\psi_2' \circ \psi_2''$ are $\pm\bar{\chi}_1$ and $\pm\bar{\psi}_2$, respectively. Using the flexibility given to us by Proposition 6.3, we can adjust the signs of the components of χ', ψ', χ'', and ψ'' so that

$$\bar{\chi}_1 = \chi_1' \circ \chi_1'' \qquad \bar{\chi}_2 = \chi_2''$$
$$\bar{\psi}_2 = \psi_2' \circ \psi_2'' \qquad \bar{\psi}_1 = \psi_1'.$$

The condition $\bar{\psi} \circ \bar{\chi} = 0$, which is equivalent to condition (iii) of Proposition 6.10, can be restated as the assertion that the exterior of the diagram above anticommutes (that is, commutes up to a minus sign). Similarly, the exactness of sequences (6.5) and (6.6) implies that the primed and double primed squares in the diagram above anticommute. It follows that, after all our sign changes have been made,

$$\psi_1'' = -\chi_2'.$$

Proposition 6.3 indicates that we can take the maps ∂' and ∂'' in sequences (6.5) and (6.6) to be any nonzero maps. The composites $\Sigma\chi_1'' \circ \bar{\partial}$ and $\bar{\partial} \circ \psi_2'$ are easily seen to be nonzero, so we take these as our choices for ∂' and ∂'', respectively.

Define maps

$$\gamma : \tilde{J} \longrightarrow J_1' \oplus \tilde{J} \oplus J_2'$$
$$s : J' = J_1' \oplus J_2' \longrightarrow J_1' \oplus \tilde{J} \oplus J_2'$$
$$\theta : J_1' \oplus \tilde{J} \oplus J_2' \longrightarrow J'$$

by the formulae

$$\begin{aligned} \gamma(x) &= (\chi_1'(x), -x, 0) \\ s(u,v) &= (u, 0, v) \\ \theta(a,b,c) &= (a + \chi_1'(b), c). \end{aligned}$$

Of course, these are maps between $RO(G)$-graded Mackey functors, so these formulae must be interpreted as applying for each $\alpha \in RO(G)$ and each of the orbits G/G and G/e.

Now consider the diagram

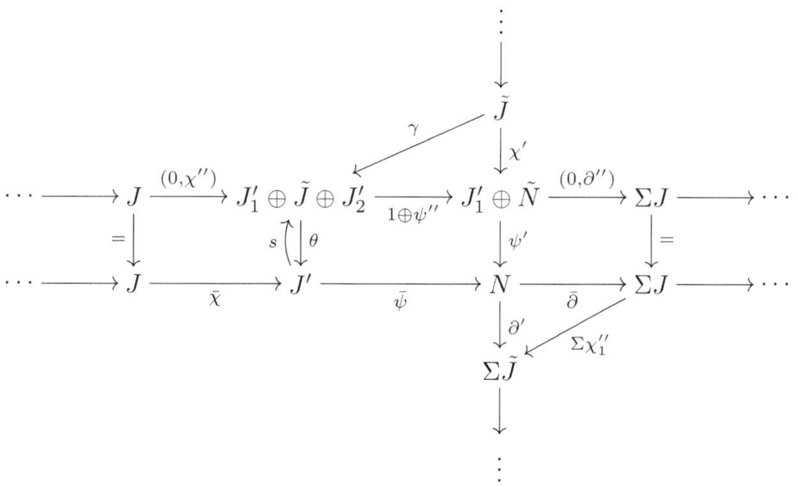

in which the vertical column is just long exact sequence (6.5). The top full row of this diagram is obtained from long exact sequence (6.6) by adding J_1' to the $\tilde{J} \oplus J_2'$ and $\tilde{N}$ terms in that sequence. Thus, the top full row is exact. The bottom row is the sequence whose exactness is to be proven. It is fairly easy to see that, if s is removed from the diagram, then the remainder of the diagram commutes. Clearly, $\theta \circ s = id$, and $\theta \circ \gamma = 0$. In fact, the two maps γ and θ form a split short exact sequence.

The two composites $\bar{\partial} \circ \bar{\psi}$ and $\Sigma\bar{\chi} \circ \bar{\partial}$ are zero by Lemma 6.1, and the composite $\bar{\psi} \circ \bar{\chi}$ is assumed to be zero. Thus, to show that the bottom row is exact, it suffices to show that the kernel of each map is contained in the image of the previous map. At N, this follows from a perfectly straightforward diagram chase. If y is an element of J'_ω, for some $\omega \in RO(G)$, and $\bar{\psi}_\omega(y) = 0$, then chasing $s(y)$ around the diagram easily gives that y is in the image of $\bar{\chi}$.

Now assume that $x \in J_\omega$, for some $\omega \in RO(G)$, and $\bar{\chi}_\omega(x) = 0$. If $(0, \chi'')_\omega(x)$ is zero, then it follows easily from the exactness of the top row of the diagram that x is in the image of $\Sigma^{-1}\bar{\partial} : \Sigma^{-1}N \longrightarrow J$. On the other hand, if $(0, \chi'')_\omega(x) \neq 0$, then there is a nonzero element z of $\tilde{J}_\omega$ such that $\gamma_\omega(z) = (0, \chi'')_\omega(x)$. Since $\chi'_\omega(z) = 0$, z must be in the image of the map $(\Sigma^{-1}\partial')_\omega : (\Sigma^{-1}N)_\omega \longrightarrow \tilde{J}_\omega$. The map $\Sigma^{-1}\partial'$ is the composite of $\Sigma^{-1}\bar{\partial}$ and χ''_1. Thus, $(\Sigma^{-1}\bar{\partial})_\omega$ must be nonzero. In any dimension where this map is nonzero, its target is either $\langle \mathbb{Z}/p \rangle$ or L_-. If the target is $\langle \mathbb{Z}/p \rangle$, then x is in the image of $(\Sigma^{-1}\bar{\partial})_\omega$ since any nonzero map into $\langle \mathbb{Z}/p \rangle$ is surjective. If the target of $(\Sigma^{-1}\bar{\partial})_\omega$ is L_-, then $p = 2$ and the domain of this map is $\langle \mathbb{Z}/2 \rangle$. Since $(\Sigma^{-1}\bar{\partial})_\omega$ is nonzero, it is surjective at G/G, and any $x \in J_\omega(G/G)$ is in its image. The map $\bar{\chi}_\omega$ is injective at G/e since $\bar{\chi}$ is constructed from standard shift maps. Thus, if $x \in J_\omega(G/e)$, then $x = 0 \in \operatorname{Im}(\Sigma^{-1}\bar{\partial})_\omega$. This completes the proof of the first stage of our induction.

For the second stage of the induction, we remove the restriction $|V - \omega_1| = 2$ from the quadruple V, ω_1, ω'_1, and ω'_2. Again, in order to apply our induction hypothesis, we want to replace this single quadruple by two others. However, this time we want these two to have smaller vertical, rather than horizontal, spreads than that of the original quadruple. Select a nontrivial complex irreducible G-representation η. Each of our new quadruples is formed by replacing a pair of the elements from the original quadruple by the elements $\omega_1 + \eta$ and $\omega'_1 + \eta$ of $RO(G)$. Thus, let $\hat{J} = \Sigma^{\omega_1+\eta}H_*$ and $\hat{N} = \Sigma^{\omega'_1+\eta}H_*$. Consider the quadruple $\omega'_1 + \eta$, ω_1, ω'_1, and $\omega_1 + \eta$ in which $\omega'_1 + \eta$ replaces V and $\omega_1 + \eta$ replaces ω'_2. Since $(\omega'_1 + \eta) + \omega_1 - \omega'_1 - (\omega_1 + \eta) = 0$, Lemma 6.5 indicates that there are maps $\chi' : J \longrightarrow J'_1 \oplus \hat{J}$ and $\psi' : J'_1 \oplus \hat{J} \longrightarrow \hat{N}$, constructed from standard shift maps, such that the composite $\psi'_{\omega_1} \circ \chi'_{\omega_1}$ is zero. Moreover, because $|(\omega'_1 + \eta) - \omega_1| = 2$, we can conclude from the first stage of our induction that the sequence

$$\cdots \longrightarrow J \xrightarrow{\chi'} J'_1 \oplus \hat{J} \xrightarrow{\psi'} \hat{N} \xrightarrow{\partial'} \Sigma J \longrightarrow \cdots . \tag{6.7}$$

is exact provided the map ∂' is nonzero.

Consider also the quadruple V, $\omega_1 + \eta$, $\omega'_1 + \eta$, and ω'_2 in which $\omega_1 + \eta$ replaces ω_1 and $\omega'_1 + \eta$ replaces ω'_1. Note that

$$V + (\omega_1 + \eta) - (\omega'_1 + \eta) - \omega'_2 = V + \omega_1 - \omega'_1 - \omega'_2.$$

Thus, by Lemma 6.5, there are maps $\chi'' : \hat{J} \longrightarrow \hat{N} \oplus J'_2$ and $\psi'' : \hat{N} \oplus J'_2 \longrightarrow N$, constructed from standard shift maps, such that the composite $\psi''_{\omega_1+\eta} \circ \chi''_{\omega_1+\eta}$ is zero. Since

$$|V - (\omega_1 + \eta)| < |V - \omega_1|,$$

our induction hypothesis for the second stage allows us to assume that the sequence

$$\cdots \longrightarrow \hat{J} \xrightarrow{\chi''} \hat{N} \oplus J'_2 \xrightarrow{\psi''} N \xrightarrow{\partial''} \Sigma \hat{J} \longrightarrow \cdots \tag{6.8}$$

is exact provided the map ∂'' is nonzero.

As in the first stage of the induction, Proposition 6.3 gives us a certain amount of flexibility in the choice of the maps χ', ψ', χ'', and ψ'' in these two long exact sequences. We want to use that flexibility to select those maps in such a way that we can derive the exactness of sequence (6.4) from the exactness of sequences (6.7) and (6.8). The six components of the maps $\bar{\chi}$, $\bar{\psi}$, χ', ψ', χ'', and ψ'' fit into the diagram

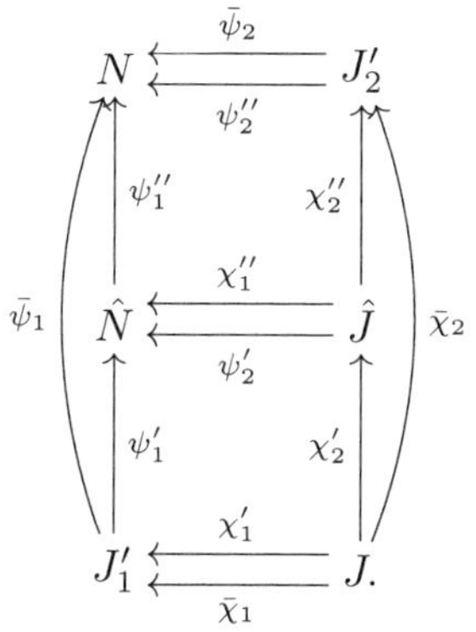

As before, all of the maps in this diagram are standard shift maps, and so the maps in each parallel pair are either equal or negatives of each other. Moreover, the composites $\chi_2'' \circ \chi_2'$ and $\psi_1'' \circ \psi_1'$ are $\pm\bar{\chi}_2$ and $\pm\bar{\psi}_1$, respectively. Using the flexibility given to us by Proposition 6.3, we can adjust the signs of the components of χ', ψ', χ'', and ψ'' so that

$$\bar{\chi}_2 = \chi_2'' \circ \chi_2' \qquad \bar{\chi}_1 = \chi_1'$$
$$\bar{\psi}_1 = \psi_1'' \circ \psi_1' \qquad \bar{\psi}_2 = \psi_2''.$$

The assumed vanishing of the composites $\bar{\psi} \circ \bar{\chi}$, $\psi' \circ \chi'$, and $\psi'' \circ \chi''$ implies that the exterior of this diagram and the primed and double primed squares in this diagram anticommute. It follows that, after all our sign adjustments have been made, $\chi_1'' = -\psi_2'$.

The only condition which the maps ∂' and ∂'' must satisfy is that they must be nonzero. It is easy to check that the composites $\bar{\partial} \circ \psi_1''$ and $\Sigma\chi_2' \circ \bar{\partial}$ are nonzero, so we take these two composites to be ∂' and ∂'', respectively.

Consider the diagram

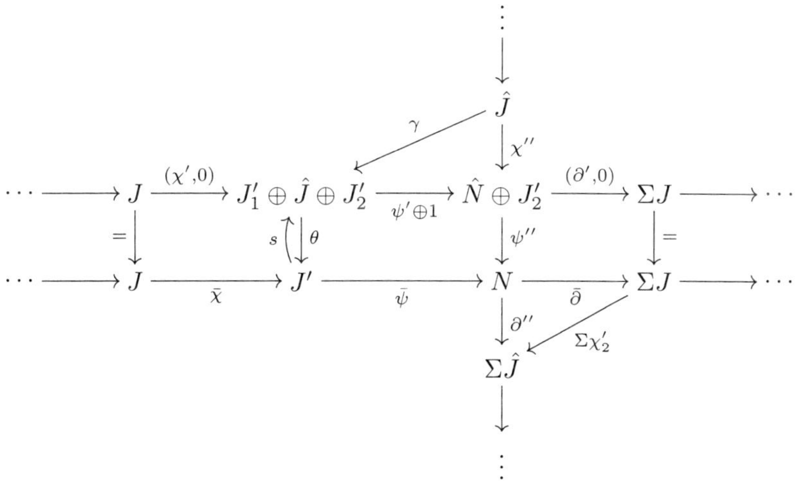

in which the maps

$$\gamma : \hat{J} \longrightarrow J_1' \oplus \hat{J} \oplus J_2'$$
$$s : J' = J_1' \oplus J_2' \longrightarrow J_1' \oplus \hat{J} \oplus J_2'$$
$$\theta : J_1' \oplus \hat{J} \oplus J_2' \longrightarrow J'$$

are defined by the formulae

$$\begin{aligned} \gamma(x) &= (0, -x, \chi_2''(x)) \\ s(u,v) &= (u, 0, v) \\ \theta(a,b,c) &= (a, \chi_2''(b) + c). \end{aligned}$$

The vertical column in this diagram is just long exact sequence (6.8). The top full row of this diagram is obtained from long exact sequence (6.7) by adding J_2' to the $J_1' \oplus \hat{J}$ and $\hat{N}$ terms in that sequence. Thus, the top full row is exact. As in the first stage of the induction, the bottom row is the sequence whose exactness is to be proven. Also, if s is removed from the diagram, then the remainder of the diagram commutes. Moreover, the two maps γ and θ form a short exact sequence which is split by s. From this point on, the argument for the exactness of the bottom row of this diagram follows exactly the same pattern as the one presented in the first stage of the induction. Thus, the proof of Proposition 6.10 is complete.

6.5. The congruence condition on $d_{(V+\sum\omega_i-\sum\omega_j')}$

In this section, we return to the context presented at the beginning of this chapter in which the complexity n of our sequences is an arbitrary positive integer. Our goal is to prove Proposition 4.9. This result describes the congruence condition on $d_{V+\sum\omega_i-\sum\omega_j'}$ that is the sole obstruction to the existence of a dimension-shifting long exact sequence associated to the elements ω_i, ω_j', and V of $RO(G)$. The information on the multiplicative structure of H_* provided by Proposition 1.10 is needed for the explicit computations required in the proof of this result.

For these computations, it is useful to define elements β_j of $RO(G)$ by

$$\beta_j = V + \sum_{i=1}^{j-1} \omega_i - \omega_i',$$

for $1 < j \leq n+1$. Observe that $|\beta_j| = |V|$ and $|\beta_j^G| = |(\omega_j')^G|$ for $1 < j \leq n+1$.

Denote the canonical generator of $J_{\omega_i} = A$ by μ_i for $1 \leq i \leq n$. The key to the proof of the assertion about $\bar{\chi}$ in Proposition 6.2 is the observation that, for any map $\bar{\chi} : J \longrightarrow J'$,

$$\bar{\chi}_{\omega_i}(\mu_i) = e_i \xi_{\omega_i - \omega_i'} + e_i' \epsilon_{\omega_i - \omega_{i+1}'}$$

for some integers e_i and e_i'. The map $\bar{\chi}$ is constructed from standard shift maps if and only if these integers are ± 1 for $1 \leq i \leq n$.

Similarly, denote the standard generator of $J'_{\omega_j'} = A$ by μ_j' for $1 \leq j \leq n+1$. From the proof of the assertion about $\bar{\psi}$ in Proposition 6.2, it follows that, for any map $\bar{\psi} : J' \longrightarrow N$, there are integers e_j'' such that

$$\bar{\psi}_{\omega_j}(\mu_j') = \begin{cases} e_1'' \epsilon_{\omega_1' - V} & \text{for } j = 1, \\ e_j'' \epsilon_{\omega_j' - \beta_j} \xi_{\beta_j - V} & \text{for } 1 < j \leq n, \\ e_{n+1}'' \xi_{\omega_{n+1}' - V} & \text{for } j = n+1. \end{cases}$$

Note that, for $1 < j \leq n$, the integer e''_j is only determined mod p. Also observe that the map $\bar{\psi}$ is constructed from standard shift maps if and only if e''_1 and e''_{n+1} are ± 1 and the e''_j are relatively prime to p for all j. These observations suffice for the computations needed in our proof.

PROOF OF PROPOSITION 4.9. We begin with the "only if" part of the proof. Assume that $\bar{\chi} : J \longrightarrow J'$ and $\bar{\psi} : J' \longrightarrow N$ are constructed from standard shift maps. From the formulae above, we obtain that

$$\begin{aligned}(\bar{\psi}_{\omega_1} \circ \bar{\chi}_{\omega_1})(\mu_1) &= e_1 e''_1 \epsilon_{\omega'_1 - V} \xi_{\omega_1 - \omega'_1} + e'_1 e''_2 \epsilon_{\omega_1 - \omega'_2} \epsilon_{\omega'_2 - \beta_2} \xi_{\beta_2 - V} \\ &= (e_1 e''_1 + e'_1 e''_2) \epsilon_{\omega'_1 - V} \xi_{\omega_1 - \omega'_1}.\end{aligned}$$

Thus, $(\bar{\psi}_{\omega_1} \circ \bar{\chi}_{\omega_1})(\mu_1) = 0$ if and only if

$$e''_2 \equiv -e_1 e'_1 e''_1 \mod p.$$

Since each of e_1, e'_1, and e''_1 is ± 1, it follows that $e''_2 \equiv \pm 1 \bmod p$ if $(\bar{\psi}_{\omega_1} \circ \bar{\chi}_{\omega_1})(\mu_1)$ is zero.

Similarly, for $1 < i < n$,

$$\begin{aligned}(\bar{\psi}_{\omega_i} \circ \bar{\chi}_{\omega_i})(\mu_i) &= e_i e''_i \epsilon_{\omega'_i - \beta_i} \xi_{\omega_i - \omega'_i} \xi_{\beta_i - V} + e'_i e''_{i+1} \epsilon_{\omega_i - \omega'_{i+1}} \epsilon_{\omega'_{i+1} - \beta_{i+1}} \xi_{\beta_{i+1} - V} \\ &= (e_i e''_i + e'_i e''_{i+1}) \epsilon_{\omega'_i - \beta_i} \xi_{\beta_{i+1} - V}.\end{aligned}$$

It follows that $(\bar{\psi}_{\omega_i} \circ \bar{\chi}_{\omega_i})(\mu_i) = 0$ if and only if

$$e''_{i+1} \equiv -e_i e'_i e''_i \mod p.$$

Inductively, this allows us to argue that $e''_{i+1} \equiv \pm 1 \bmod p$ if $(\bar{\psi}_{\omega_j} \circ \bar{\chi}_{\omega_j})(\mu_j) = 0$ for $j \leq i$.

Finally,

$$\begin{aligned}(\bar{\psi}_{\omega_n} \circ \bar{\chi}_{\omega_n})(\mu_n) &= e_n e''_n \epsilon_{\omega'_n - \beta_n} \xi_{\omega_n - \omega'_n} \xi_{\beta_n - V} + e'_n e''_{n+1} \epsilon_{\omega_n - \omega'_{n+1}} \xi_{\omega'_{n+1} - V} \\ &= (e_n e''_n + d_{\beta_{n+1} - \omega'_{n+1}} e'_n e''_{n+1}) \epsilon_{\omega_n - \beta_{n+1}} \xi_{\beta_{n+1} - V}.\end{aligned}$$

This gives that $(\bar{\psi}_{\omega_n} \circ \bar{\chi}_{\omega_n})(\mu_n) = 0$ if and only if

$$d_{\beta_{n+1} - \omega'_{n+1}} e''_{n+1} \equiv -e_n e'_n e''_n \mod p.$$

Since $e''_{n+1} = \pm 1$, we can conclude that $d_{\beta_{n+1} - \omega'_{n+1}} \equiv \pm 1 \bmod p$ if $(\bar{\psi}_{\omega_i} \circ \bar{\chi}_{\omega_i})(\mu_i)$ is zero for $1 \leq i \leq n$. The observation that

$$\beta_{n+1} - \omega'_{n+1} = V + \sum_{1 \leq i \leq n} \omega_i - \sum_{1 \leq j \leq n+1} \omega'_j,$$

completes the "only if" part of the proof of the proposition.

For the "if" part, assume that integers e_i and e'_i, for $1 \leq i \leq n$, and an integer e''_1 have been chosen so that each is ± 1. If $d_{\beta_{n+1} - \omega'_{n+1}} \equiv \pm 1 \mod p$, then we can select integers $e''_i = \pm 1$, for $1 < i \leq n+1$, which satisfy the appropriate congruences noted above. This collection of integers specifies maps $\bar{\chi} : J \longrightarrow J'$ and $\bar{\psi} : J' \longrightarrow N$ such that $\bar{\psi}_{\omega_i} \circ \bar{\chi}_{\omega_i} = 0$ for all i. □

CHAPTER 7

Complex Grassmannian manifolds

If V is a complex G-representation and k is a positive integer, then the Grassmannian manifold $G(V,k)$ of complex k-dimensional subspaces of V carries an obvious G-action derived from the action of G on V. Nonequivariantly, $G(V,k)$ is a CW-complex whose cells are the Schubert cells (see, for example, [**8, 9, 20**]). Here, we show that, if G is a finite abelian group, then there is an equivariant version of the Schubert cell structure of $G(V,k)$ which provides this G-space with the structure of a $\mathrm{Rep}^*(G)$-cell complex. We also show that, for $G = \mathbb{Z}/p$, this cell structure on $G(V,k)$ satisfies the finite type hypothesis of Corollary 2.8 so that the equivariant $RO(G)$-graded ordinary homology $H_*^G(G(V,k);A)$ is free over H_*. These results are presented in the first section of this chapter. In the second section, we examine in detail the Grassmannian manifold of complex 3-dimensional subspaces of a 6-dimensional complex $\mathbb{Z}/p$-representation V in order to illustrate the computational difficulties caused by the dimension shifting inherent in our freeness results.

7.1. Equivariant Schubert cells and $H_*^G(G(V,k);A)$

To describe the $\mathrm{Rep}^*(G)$-cell structure on $G(V,k)$, we assume initially that V is a finite dimensional complex G-representation and express $V = \bigoplus_{s=1}^{m} \phi_s$ as a sum of complex irreducibles. Since we are assuming that G is abelian, these irreducible representations ϕ_s have complex dimension one. From this description of V in terms of an ordered collection of irreducible representations, we obtain a flag of subspaces

$$0 < V_1 < V_2 < \cdots < V_m = V$$

of V in which $V_t = \bigoplus_{s=1}^{t} \phi_s$ for each $1 \le t \le m$. In terms of this fixed flag, the standard Schubert cells can now be described as usual. Here, we follow the notation used in [**9**]. Given a sequence of integers $0 \le a_1 \le \ldots \le a_k \le m-k$, define the cell $\langle a_1, \ldots, a_k \rangle$ by

$$\langle a_1, \ldots, a_k \rangle = \{X \in G(V,k) : \dim_{\mathbb{C}}(X \cap V_{a_i+i}) = i \text{ for all } i \text{ such that } 1 \le i \le k\}.$$

This cell $\langle a_1, \ldots, a_k \rangle$ is usually represented as a matrix

$$\begin{array}{cccccccccccccc}
\phi_1 & \cdots & \phi_{a_1} & \phi_{a_1+1} & & \cdots & & \phi_{a_2+2} & & \cdots & \phi_{a_k+k} & & \cdots & \phi_m
\end{array}$$
$$\begin{bmatrix}
* & \cdots & * & 1 & 0 & \cdots & & & & & & & & \\
* & \cdots & * & 0 & * & \cdots & * & 1 & 0 & \cdots & & & & \\
\vdots & & & & & & & & & \ddots & & & & \\
* & \cdots & * & 0 & * & \cdots & * & 0 & * & \cdots & 1 & 0 & \cdots & 0
\end{bmatrix}$$

written in standard form. Here, $*$ is used to denote an arbitrary complex number, and there are a_i $*$'s in the ith row, for each $1 \le i \le k$. The G-action on the cell $\langle a_1, \dots, a_k \rangle$ comes from letting G act on each entry of the matrix as it acts on the irreducible representation above its column. The cell $\langle a_1, \dots, a_k \rangle$ is the interior of the representation cell DW where

$$W = \bigoplus_{i=1}^{k} \bigoplus_{\substack{j=1 \\ j \notin \{a_1+1, \dots, a_{i-1}+i-1\}}}^{a_i+i-1} \phi_{a_i+i}^{-1}\phi_j.$$

Here, $\phi_{a_i+i}^{-1}\phi_j$ denotes the tensor product over $\mathbb{C}$ of ϕ_j and the conjugate of ϕ_{a_i+i}. Observe that the (real) dimension of the cell $\langle a_1, \dots, a_k \rangle$ is $|W| = 2\Sigma_{i=1}^k a_i$.

The space $G(V,k)$ is built from these cells by beginning with the 0-cell $\langle 0, \dots, 0 \rangle$ as the 0-filtration X_0. The cell $\langle a_1, \dots, a_k \rangle$ can be attached to any subcomplex of $G(V,k)$ containing all the cells $\langle b_1, \dots, b_k \rangle$ such that $b_i \le a_i$, for $1 \le i \le k$, and $b_i < a_i$ for at least one i. The usual filtration of $G(V,k)$ is specified by the equation

$$X_n - X_{n-1} = \{\langle a_1, \dots, a_k \rangle | \sum_i a_i = n\}.$$

However, there are other useful filtrations. If $V \subset V'$, then clearly $G(V,k)$ is a sub-G-cell complex of $G(V',k)$. Hence, $G(V,k)$ carries an obvious G-cell structure if V is a countably infinite dimensional complex representation.

If $G = \mathbb{Z}/p$ and V is finite dimensional, then Theorem 2.1 obviously implies that the $RO(G)$-graded homology $H_*^G(G(V,k);A)$ of $G(V,k)$ is free over H_*. However, if V is countably infinite dimensional, then Corollary 2.8 must be used to show that the homology is free. In order to apply that result, we need to find a lower bound for the fixed dimension of the representation W associated to the cell $\langle a_1, \dots, a_k \rangle$. Note that W contains one copy of the irreducible trivial complex G-representation for each pair of positive integers i and j such that $i \le k$, $j \le a_i + i - 1$, $j \ne a_t + t$ for any $t < i$, and ϕ_j is isomorphic to ϕ_{a_i+i} as a complex G-representation. In order to obtain a useful lower bound for the number of such pairs, we need to choose the ordering on the irreducible summands ϕ_s of V carefully. For each positive integer r, let B_r be a set of representatives of the isomorphism classes of irreducible complex G-representations that appear in the decomposition of V at least r times. Note that, if the dimension of V is countable, then the disjoint union $\cup_r B_r$ can be identified with the set of irreducible summands of V. Assign each set B_r some ordering, and then order the set $\cup_r B_r$ as the concatenation $B_1, B_2, \dots, B_r, \dots$ of the ordered sets B_r. Each of the B_r is referred to as a block, and an ordering of the irreducible summands of V constructed in this fashion is referred to as a standard block ordering of the cells. For each positive integer n, let V_n be the sum of the first n irreducible summands of V in a standard block ordering of these summands. The sequence of inclusions

$$* = G(V_k,k) \subset G(V_{k+1},k) \subset \dots \subset G(V_n,k) \subset \dots$$

provides a filtration of $G(V,k)$ by finite sub-G-cell complexes. The control over the fixed dimensions of the cells of $G(V,k)$ needed to apply Corollary 2.8 is provided by the following result:

PROPOSITION 7.1. *Let $G = \mathbb{Z}/p$, and let V be a countably infinite dimensional complex G-representation. Assume that the irreducible complex summands of V*

have been given a standard block ordering. Then, for any positive integer m, every cell of $G(V,k)$ having fixed dimension at most $2m$ is contained in the finite subcomplex $G(V_{(m+k)p},k)$ of $G(V,k)$.

PROOF. Observe that the cell $\langle a_1,\ldots,a_k\rangle$ of $G(V,k)$ is contained in the subcomplex $G(V_n,k)$ of $G(V,k)$ if and only if $a_k+k\le n$. Assume that the cell $\langle a_1,\ldots,a_k\rangle$ of $G(V,k)$ is added after $G(V_{(m+k)p},k)$ so that $a_k+k>(m+k)p$. Since there are only p distinct irreducible complex representations of G, irreducible representations isomorphic to ϕ_{a_k+k} must have appeared at least $m+k$ times prior to ϕ_{a_k+k} in our standard block ordering of the irreducible summands of V. At most $k-1$ of these appearances can come from the set $\{\phi_{a_i+i}|1\le i<k\}$. Hence, the irreducible trivial complex G-representation must appear at least $m+1$ times in the portion of the sum

$$W=\bigoplus_{i=1}^{k}\bigoplus_{\substack{j=1\\ j\notin\{a_1+1,\ldots,a_{i-1}+i-1\}}}^{a_i+i-1}\phi_{a_i+i}^{-1}\phi_j$$

coming from $i=k$. Thus, the fixed dimension of the representation W associated to the cell $\langle a_1,\ldots,a_k\rangle$ is at least $2(m+1)$. □

Since, for any positive integer m, there are only finitely many cells of $G(V,k)$ having fixed dimension below $2m$, Corollary 2.8 applies to $G(V,k)$.

COROLLARY 7.2. *Let $G=\mathbb{Z}/p$. If V is a finite or countably infinite dimensional complex G-representation, then the $RO(G)$-graded Mackey functor-valued equivariant ordinary homology $H_*^G(G(V,k);A)$ of $G(V,k)$ is free over H_*.*

In this corollary, as in the results from Chapter 2 on which it is based, nothing is said about the dimension of the generators of the homology. For spaces as well-behaved as Grassmannian manifolds, this omission is somewhat embarrassing. However, the example presented in the next section indicates that it is still necessary.

7.2. A calculational example

The dimensions of the generators for the cohomology of complex projective spaces are given in [**12**], and those results apply equally well in homology. There is a plausible conjecture for the dimensions of the generators of $H_*^G(G(V,2);A)$ for any G-representation V. However, the only obvious route to verifying that conjecture would require a tedious argument. For $k\ge 3$ and representations V with complex dimension at least 6, there is no credible conjecture about the dimensions of the generators. For such k and V, except for a few special cases, it seems impossible to select the Schubert cells in a way that eliminates the possibility of dimension shifting. The following example illustrates the sort of difficulties that can arise.

Let λ_1 and λ_2 be two non-isomorphic irreducible complex representations of $G=\mathbb{Z}/p$, and let V be the sum of three copies of each of λ_1 and λ_2. Here we consider the problem of determining the dimensions of $H_*^G(G(V,3);A)$. These are the smallest k and V for which there is no credible conjecture about the dimensions of those generators. Note that our choice of having exactly two nonisomorphic irreducibles appear in V is the worst possible case. If only one irreducible appeared, then the G-action would be trivial and the homology would be obvious. Having a larger number of distinct irreducibles present reduces the size of the fixed point subset and so reduces the number of dimension shifts that might occur. For example, if V were, instead, the sum of six non-isomorphic irreducible complex representations, then no dimension-shifting could occur, and the homology is easily computed.

In order to specify the equivariant Schubert cells of $G(V, 3)$, we must order the irreducible summands of V. As is pointed out in [**12**], to reduce the problems with nonvanishing boundary maps in the cell-attaching long exact sequences, it is important to ensure that the low dimensional cells of the space have fixed point dimensions as small as possible. This can be accomplished by ordering the irreducibles in such a way that the repeated appearances of each isomorphism class of irreducibles are as widely separated as possible. Thus, the order λ_1, λ_2, λ_1, λ_2, λ_1, λ_2 is one of two appropriate choices for this example. Figure 7.1 contains a plot of the dimensions (regarded as elements of $RO(G)$) of the Schubert cells for $G(V, 3)$ produced by this ordering. In this plot, ■ is used to denote a single cell plotting to a particular location. If more than one cell plots to a location, then the number of cells plotting to that location is placed there.

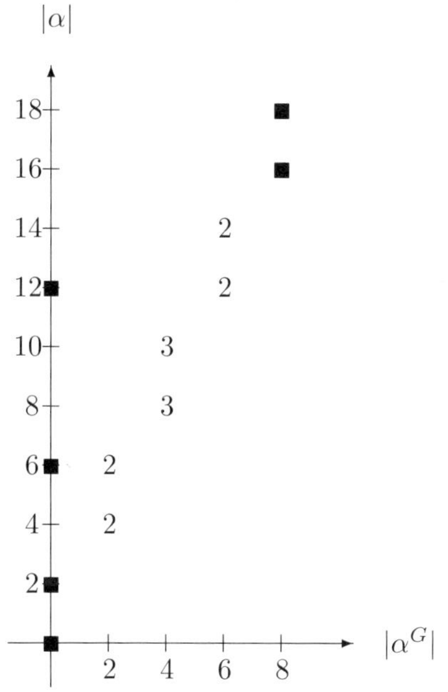

FIGURE 7.1. The Schubert Cells of $G(V, 3)$ with the ordering λ_1, λ_2, λ_1, λ_2, λ_1, λ_2

It should be clear from the figure that attaching the cells whose dimensions plot to $(0, 6)$ and $(0, 12)$ might cause dimension shifting. A shift caused by the cell plotting to $(0, 6)$ could only effect the generator associated to one of the two cells plotting to $(2, 4)$. It can be argued that this shift must, in fact, occur. The cell plotting to $(0, 12)$ creates more difficulties. It might shift the dimension of only one generator. That single generator could be the one associated to a cell plotting to any of the points $(2, 4)$, $(2, 6)$, $(4, 8)$, or $(4, 10)$. It could also be the one associated to the cell at $(0, 6)$ since that cell causes a dimension shift and so its generator plots to $(2, 6)$. Alternatively, the cell at $(0, 12)$ could shift the dimensions of two generators.

One of those two would have to be associated to a cell plotting at either $(4,8)$ or $(4,10)$. The other could be either the generator associated to a cell plotting at at $(2,4)$ or $(2,6)$ or the shifted generator associated to the cell at $(0,6)$. Using only this one Schubert cell structure for $G(V,3)$, the only way to resolve the question of which of these shifts, if any, occurs would be to explicitly compute the boundary map in the cell attaching long exact sequence. That computation appears to be nontrivial.

There is, however, another approach to resolving this problem. For $1 \leq t \leq 6$, let V_t be the sum of the first t irreducibles in the sequence λ_1, λ_2, λ_1, λ_2, λ_1, λ_2, so that $V = V_6$. In the process of obtaining this Schubert cell structure for $G(V,3)$, we also produce Schubert cell structures for $G(V_4,3)$ and $G(V_5,3)$. Keeping this cell structure for $G(V_4,3)$, we could select a different collection of cells to add to $G(V_4,3)$ to produce $G(V_5,3)$ by reordering the first 4 irreducibles in our ordered list λ_1, λ_2, λ_1, λ_2, λ_1, λ_2. For example, we could produce the cells of $G(V_5,3)-G(V_4,3)$ using the reordering λ_1, λ_1, λ_2, λ_2, λ_1, λ_2. From the viewpoint of nonequivariant homotopy theory, this is a rather perverse thing to do. However, in the equivariant context, it produces a set of cells for $G(V_5,3)$ which have different fixed point dimensions and so different potential dimension shifting problems. The dimensions of the cells for $G(V_5,3)$ with both our original ordering and this shifted ordering are displayed in Figure 7.2.

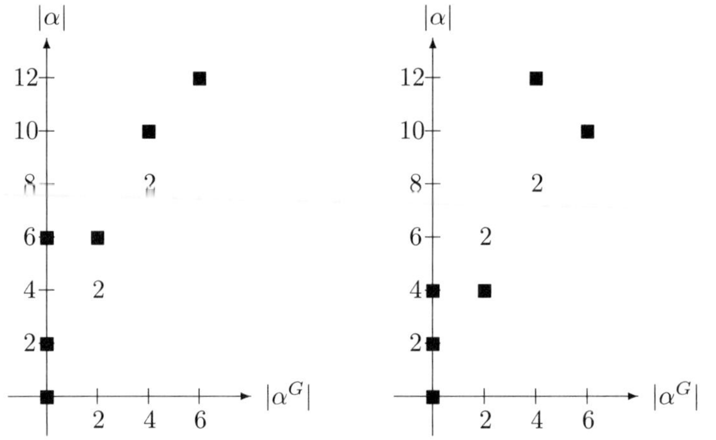

FIGURE 7.2. The Schubert Cells of $G(V_5,3)$: Original ordering on the left, shifting after $G(V_4,3)$ on the right

The only dimension-shifting that can occur for this new cell structure on $G(V_5,3)$ is that the cell plotting to $(4,12)$ could shift the generator associated to the cell plotting to $(6,10)$. Since these two cell structures for $G(V_5,3)$ must yield the same homology for $G(V_5,3)$, it is easy to see that this potential shift must, in fact, occur. The same sort of argument indicates, as we suggested earlier, that the cell plotting to $(0,6)$ in the original cell structure must produce a dimension-shift involving the generator associated to one of the two cells plotting to $(2,4)$ in that original structure. These arguments indicate that the generators of $H_*^G(G(V_5,3);A)$ must appear in the dimensions plotted in Figure 7.3.

We have, unfortunately, still not completely determined the dimensions of the generators of $H_*^G(G(V_5,3);A)$. When a cell attachment forces a generator to shift

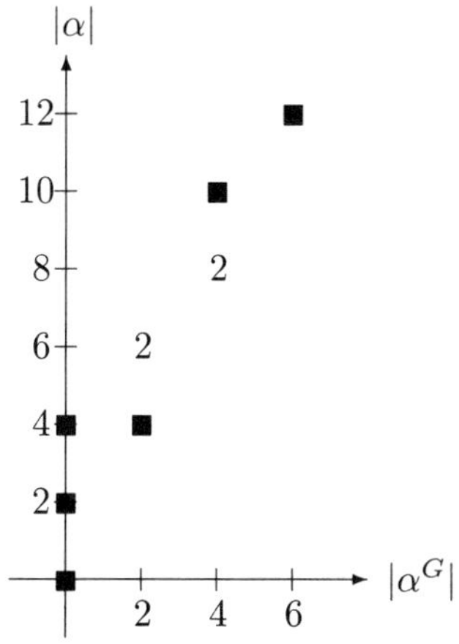

FIGURE 7.3. The dimensions of the generators of $H_*^G(G(V_5, 3); A)$

to a new dimension ω, the integers $|\omega^G|$ and $|\omega|$ associated to that new dimension can be obtained from the stairstep pattern associated to that cell attachment. The proof of our freeness result offers no help in deciding which of the elements of $RO(G)$ plotting to that point is the actual dimension of the generator. Corollary 4.11 is our only source of information about the actual value of ω, as opposed to the integers $|\omega^G|$ and $|\omega|$. However, since it deals only with the integer d associated to a linear combination of the dimensions of all the generators involved in the shift, it is of very little value if the ramp of generators being shifted is a long one. In the particular case of $H_*^G(G(V_5, 3); A)$, ad hoc arguments or the even more unorthodox cell structure described below, can be used to determine the actual dimensions of the generators as opposed to the points at which they plot.

Note that $G(V_5, 3)$ is G-homeomorphic to $G(V_5, 2)$. An extended version of the sort of comparison of multiple cell structures carried out here is the source of the conjecture on the dimensions of the generators for $H_*^G(G(V, 2); A)$ for an arbitrary V. The difficulties in filling in the details of this argument for a large representation V containing a sizable number of distinct irreducibles should be easy to imagine.

We return now to the problem of determining the dimensions of the generators of $H_*^G(G(V, 3); A)$. Having chosen a cell structure for $G(V_5, 2)$, we can reorder the first five irreducibles in our ordered list λ_1, λ_2, λ_1, λ_2, λ_1, λ_2. The optimal reordering seems to be λ_2, λ_2, λ_1, λ_1, λ_1, λ_2. By plotting the cells of $G(V, 3) - G(V_5, 3)$ produced by this reordering together with the generators of $H_*^G(G(V_5, 3); A)$, one can see that three of the new cells might cause dimension shifts. Two of these questions about dimension shifts are easily resolved by comparing the new cell structure to the original one, as we did in analyzing $G(V_5, 3)$. However, the third potential shift cannot be resolved completely because it affects some of the cells that are involved in the potential dimension shift caused by the cell at $(0, 12)$ in the original structure. One can determine that the cell plotting to $(0, 12)$ in the original structure must cause a dimension shift involving a cell plotting to $(4, 8)$ and a cell plotting to either $(2, 4)$ or $(2, 6)$ in that original structure. We have not been able to determine which of those two cells is involved in this shift.

The trick of reordering the list of irreducible representations to produce alternative cell structures can be refined somewhat. In the process of constructing

$G(V_5, 2)$ from $G(V_4, 2)$, adding the cells $\langle 0, 0, 2\rangle$, $\langle 0, 1, 2\rangle$, and $\langle 1, 1, 2\rangle$ brings us to a point at which it is possible to reorder the first three irreducibles in our list before selecting the three additional cells that must be added to complete the construction of $G(V_5, 2)$. Reordering the irreducibles back to the original sequence λ_1, λ_2, λ_1, λ_2, λ_1, λ_2 at this point produces a cell structure for $G(V_5, 2)$ whose dimensions plot to exactly the same places as the generators of $H_*^G(G(V_5, 3); A)$ displayed in Figure 7.3. With this new cell structure for $G(V_5, 3)$, there are no potential dimension shifts, and so the actual dimensions of the generators are the same as the dimensions of the new cells. In the process of constructing $G(V, 3)$ from $G(V_5, 3)$, there are two intermediate points at which it is possible to reorder some portion of the list of generators. By a careful choice of these reorderings, one can reduce the number of cells that might cause dimension shifts down to one. Unfortunately, the question of whether that cell actually causes a dimension shift is equivalent to our earlier question, for the first cell structure on $G(V, 3)$, about the dimension of the second cell involved in the shift caused by the cell at $(0, 12)$. The plot on the left in Figure 7.4 displays the dimensions of the cells for this optimally chosen cell structure for $G(V, 3)$. The unresolved dimension-shifting question arising in that structure is whether the cell plotting to $(0, 6)$ causes a shift involving the generator associated to one of the two cells plotting to $(2, 4)$. The two plots in Figure 7.4 show the two possible sets of dimensions of the generators of $H_*^G(G(V, 3); A)$.

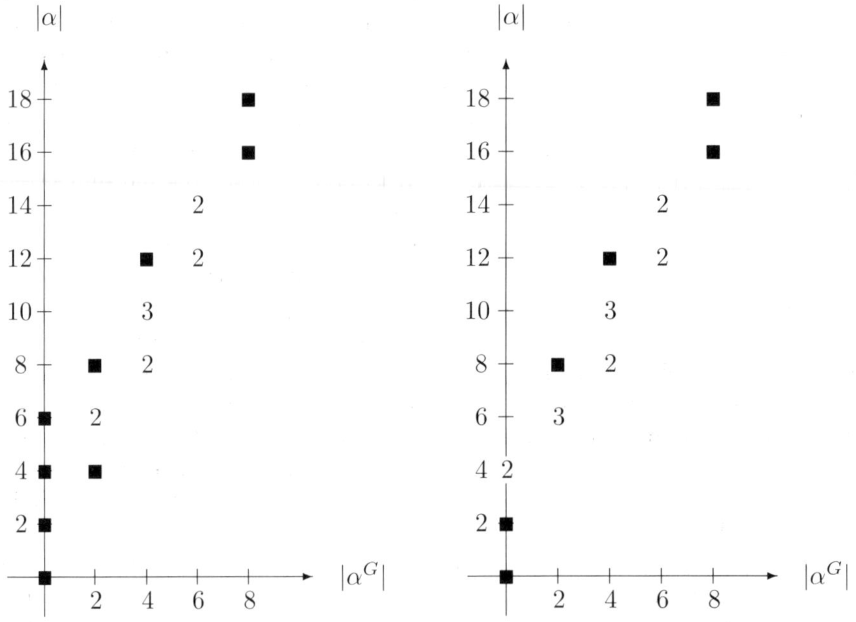

FIGURE 7.4. The two possible sets of dimensions for the generators of $H_*^G(G(V, 3); A)$: Unshifted on the left, shifted on the right. The left plot also shows the dimensions of the cells for the optimal cell structure

One real difficulty with the unorthodox cell structures we have described here must be mentioned. For the standard Schubert cell structure, there is a very simple algorithm for deciding when a cell $\langle a_1, \ldots, a_k\rangle$ is attached to a cell $\langle b_1, \ldots, b_k\rangle$. That algorithm no longer works if one reorders the list of irreducibles at some stage of constructing the cells. With the reordering, it is still possible to decide when one cell is attached to another. However, it is a somewhat more difficult.

Part 2

Observations about $RO(G)$-graded equivariant ordinary homology

CHAPTER 8

The computation of H_*^S for arbitrary S

Here, we assume that $G = \mathbb{Z}/p$ and compute the $RO(G)$-graded equivariant ordinary homology H_*^S of a point with an arbitrary Mackey functor S as coefficients. The approach taken is a variant of that employed in [**6**] and [**12**]. Our basic tools are derived from unpublished work of Stong.

In order to describe H_*^S, we need to introduce some new maps and a collection of new Mackey functors. The restriction map $\rho : S(G/G) \longrightarrow S(G/e)$ and transfer map $\tau : S(G/e) \longrightarrow S(G/G)$ of the Mackey functor S induce maps

$$\tilde{\rho} : S(G/G) \longrightarrow S(G/e)^G \quad \text{and} \quad \tilde{\tau} : S(G/e)/G \longrightarrow S(G/G).$$

They also induce the maps of Mackey functors

$$\widehat{\rho}_S : S \longrightarrow S_{G/e} \quad \text{and} \quad \widehat{\tau}_S : S_{G/e} \longrightarrow S$$

described in Section 1.1. The cokernel of $\widehat{\rho}_S$ and the kernel of $\widehat{\tau}_S$ play an important role in our computations, and so are denoted $\mathcal{C}(S)$ and $\mathcal{K}(S)$, respectively.

For any abelian group B carrying a G-action, there is a trace map $tr : B \longrightarrow B$ given on $x \in B$ by $tr(x) = \sum_{g \in G} gx$. This map induces maps $B/G \longrightarrow B$ and $B \longrightarrow B^G$, which we also denote by tr. The Mackey functors L and R introduced in Section 1.1 are special cases of two general constructions which produce Mackey functors $L(B)$ and $R(B)$ from any $\mathbb{Z}[G]$-module B. Diagramatically, these Mackey functors are given by:

$$L(B) \qquad\qquad\qquad\qquad R(B)$$

$$\begin{array}{ccc} B/G & \qquad\qquad & B^G \\ tr \downarrow \uparrow \pi & & i \downarrow \uparrow tr \\ B & & B \\ \circlearrowleft \theta & & \circlearrowleft \theta \end{array}$$

Here, π and i denote the projection onto the orbit module and the inclusion of the fixed point submodule, respectively, and θ denotes the action of G on B. In [**15**], these constructions are denoted $\mathcal{L}_e B$ and $\mathcal{J}_e B$, respectively. There it is shown that these functors are the left and right adjoint, respectively, to the functor from the category $\mathcal{M}$ of Mackey functors to the category of $\mathbb{Z}[G]$-modules which sends a Mackey functor S to $S(G/e)$.

In the remainder of this chapter, we often need to consider the Mackey functors $L(S(G/e))$ and $R(S(G/e))$ derived from a Mackey functor S. For compactness of notation, we denote these by $\mathfrak{L}(S)$ and $\mathfrak{R}(S)$, respectively. Observe that each of S, $\mathfrak{L}(S)$, and $\mathfrak{R}(S)$ has value $S(G/e)$ at G/e. The counit and unit of the appropriate adjunctions give us canonical maps

$$\bar{\epsilon}_S : \mathfrak{L}(S) \longrightarrow S \quad \text{and} \quad \bar{\eta}_S : S \longrightarrow \mathfrak{R}(S).$$

These two maps are the unique maps whose value at G/e is the identity map.

If $p = 2$, then there is a sign action of G on $\mathbb{Z}$ which sends 1 to -1. Denote $\mathbb{Z}$ with this action by $\mathbb{Z}_-$. The Mackey functors L_- and R_- introduced in Section 1.1 are just $L(\mathbb{Z}_-)$ and $B(\mathbb{Z}_-)$, respectively. Just as $L(B)$ and $R(B)$ are generalizations of L and R, the Mackey functors $L(B \otimes \mathbb{Z}_-)$ and $R(B \otimes \mathbb{Z}_-)$, which we denote by $L_-(B)$ and $R_-(B)$, generalize L_- and R_-. For any $\mathbb{Z}/2$-Mackey functor S, $L_-(S(G/e))$ and $R_-(S(G/e))$ are denoted $\mathfrak{L}_-(S)$ and $\mathfrak{R}_-(S)$, respectively.

Typically, the Mackey functor H_ω^S depends only on the integers $|\omega|$ and $|\omega^G|$. Thus, H_*^S is most easily visualized by plotting it out in the plane. The one case in which the integers $|\omega|$ and $|\omega^G|$ don't suffice to determine H_ω^S is that in which $\omega \in RO_0(G)$; that is, when $|\omega| = |\omega^G| = 0$. For such an ω, the value of d_ω is the only additional bit of information needed to determine H_ω^S.

THEOREM 8.1. *Let $\omega \in RO(G)$. Then,*

(i) *if $|\omega^G| = 0$,*

$$H_\omega^S = \begin{cases} \operatorname{Ker}\left(\widehat{\rho}_S : S \longrightarrow S_{G/e}\right) & \text{if } |\omega| > 0, \\ A[d_\omega] \,\square\, S & \text{if } |\omega| = 0, \\ \operatorname{Coker}\left(\widehat{\tau}_S : S_{G/e} \longrightarrow S\right) & \text{if } |\omega| < 0. \end{cases}$$

(ii) *if $|\omega^G| > 0$ and $|\omega| = 0$,*

$$H_\omega^S = \begin{cases} \mathfrak{R}(S) & \text{if } |\omega^G| \text{ is even}, \\ \mathfrak{R}_-(S) & \text{if } |\omega^G| \geq 3 \text{ and odd}, \\ \mathcal{K}(S) & \text{if } |\omega^G| = 1. \end{cases}$$

(iii) *if $|\omega^G| < 0$ and $|\omega| = 0$,*

$$H_\omega^S = \begin{cases} \mathfrak{L}(S) & \text{if } |\omega^G| \text{ is even}, \\ \mathfrak{L}_-(S) & \text{if } |\omega^G| \leq -3 \text{ and odd}, \\ \mathcal{C}(S) & \text{if } |\omega^G| = -1. \end{cases}$$

(iv) *if $|\omega^G|$ and $|\omega|$ are both positive or both negative, then $H_\omega^S = 0$.*

(v) *if $|\omega^G| > 0$ and $|\omega| < 0$,*

$$H_\omega^S = \begin{cases} \operatorname{Coker}\left(\bar{\eta}_S \circ \bar{\epsilon}_S : \mathfrak{L}(S) \longrightarrow \mathfrak{R}(S)\right) & \text{if } |\omega^G| \text{ is even}, \\ \operatorname{Ker}\left(\bar{\eta}_S \circ \bar{\epsilon}_S : \mathfrak{L}(S) \longrightarrow \mathfrak{R}(S)\right) & \text{if } |\omega^G| \geq 3 \text{ and odd}, \\ \operatorname{Ker}\left(\bar{\epsilon}_S : \mathfrak{L}(S) \longrightarrow S\right) & \text{if } |\omega^G| = 1. \end{cases}$$

(vi) *if $|\omega^G| < 0$ and $|\omega| > 0$,*

$$H_\omega^S = \begin{cases} \operatorname{Ker}\left(\bar{\eta}_S \circ \bar{\epsilon}_S : \mathfrak{L}(S) \longrightarrow \mathfrak{R}(S)\right) & \text{if } |\omega^G| \text{ is even}, \\ \operatorname{Coker}\left(\bar{\eta}_S \circ \bar{\epsilon}_S : \mathfrak{L}(S) \longrightarrow \mathfrak{R}(S)\right) & \text{if } |\omega^G| \leq -3 \text{ and odd}, \\ \operatorname{Coker}\left(\bar{\eta}_S : S \longrightarrow \mathfrak{R}(S)\right) & \text{if } |\omega^G| = -1. \end{cases}$$

REMARK 8.2. (a) If $|\omega| \neq 0$, then H_ω^S vanishes at G/e because the nonequivariant spectrum underlying the equivariant Eilenberg-Mac Lane spectrum representing homology with S-coefficients is the nonequivariant Eilenberg-Mac Lane spectrum associated to $S(G/e)$.

(b) In parts (ii) and (iii) of Theorem 8.1, the second and third cases occur only if $p = 2$.

(c) In the proof of this theorem, we show that $\operatorname{Ker}\big(\bar{\eta}_S \circ \bar{\epsilon}_S : \mathfrak{L}(S) \longrightarrow \mathfrak{R}(S)\big)$ can also be described as $\operatorname{Ker}\big(\widehat{\rho}_{\mathfrak{L}(S)} : \mathfrak{L}(S) \longrightarrow \mathfrak{L}(S)_{G/e}\big)$. Similarly, the two maps $\bar{\eta}_S \circ \bar{\epsilon}_S : \mathfrak{L}(S) \longrightarrow \mathfrak{R}(S)$ and $\widehat{\tau}_{\mathfrak{R}(S)} : \mathfrak{R}(S)_{G/e} \longrightarrow \mathfrak{R}(S)$ have isomorphic cokernels.

(d) For $p = 2$, alternative descriptions for the values of H_ω^S described in parts (v) and (vi) of the theorem are given in Proposition 8.9 and Remark 8.10. It is fairly easy to see that these alternative descriptions actually yield the same Mackey functors.

Six of the values of H_*^S appearing in the theorem above vanish for some common choices of S.

LEMMA 8.3. (a) *If the restriction map $\rho : S(G/G) \longrightarrow S(G/e)$ is a monomorphism, then $H_\omega^S = 0$ for those $\omega \in RO(G)$ such that $|\omega^G| = 0$ and $|\omega| > 0$.*

(b) *If the transfer $\tau : S(G/e) \longrightarrow S(G/G)$ is an epimorphism, then $H_\omega^S = 0$ for those $\omega \in RO(G)$ such that $|\omega^G| = 0$ and $|\omega| < 0$.*

(c) *If the transfer $\tau : S(G/e) \longrightarrow S(G/G)$ is a monomorphism, then $H_\omega^S = 0$ for those $\omega \in RO(G)$ such that $|\omega^G| = 1$ and $|\omega| < 0$.*

(d) *If the map $\tilde{\rho} : S(G/G) \longrightarrow S(G/e)^G$ is an epimorphism, then $H_\omega^S = 0$ for those $\omega \in RO(G)$ such that $|\omega^G| = -1$ and $|\omega| > 0$.*

(e) *If the restriction map $\rho : \mathfrak{L}(S)(G/G) \longrightarrow \mathfrak{L}(S)(G/e)$ is a monomorphism, then $H_\omega^S = 0$ for those $\omega \in RO(G)$ such that either $|\omega^G|$ is odd and $|\omega| < 0$ or $|\omega^G|$ is even and $|\omega| > 0$.*

PROOF. Parts (a) and (b) follow easily from the fact that, at G/G, the maps $\widehat{\rho}_S : S \longrightarrow S_{G/e}$ and $\widehat{\tau}_S : S_{G/e} \longrightarrow S$ can be identified with the restriction map $\rho : S(G/G) \longrightarrow S(G/e)$ and the transfer map $\tau : S(G/e) \longrightarrow S(G/G)$, respectively. Part (c) follows from the fact that, at G/G, the map $\bar{\epsilon}_S : \mathfrak{L}(S) \longrightarrow S$ is just the map $\tilde{\tau} : S(G/e)/G \longrightarrow S(G/G)$. If the map $\tau : S(G/e) \longrightarrow S(G/G)$ is a monomorphism, then G must act trivially on $S(G/e)$. Thus, $S(G/e)/G = S(G/e)$, and $\tilde{\tau} = \tau$. It follows that $\operatorname{Ker}\big(\bar{\epsilon}_S : \mathfrak{L}(S) \longrightarrow S\big) = 0$. Part (d) follows directly from the fact that, at G/G, the map $\bar{\eta}_S : S \longrightarrow \mathfrak{R}(S)$ is just the map $\tilde{\rho} : S(G/G) \longrightarrow S(G/e)^G$.

If $\omega \in RO(G)$ satisfies the conditions indicated in part (e), then either $H_\omega^S = 0$ by part (iv) of Theorem 8.1 or H_ω^S is the kernel of a map of Mackey functors whose domain is $\mathfrak{L}(S)$ and whose value at G/e is the identity map of $S(G/e)$. The asserted vanishing therefore follows from the observation that a map $f : M \longrightarrow N$ of Mackey functors is a monomorphism if the maps $\rho : M(G/G) \longrightarrow M(G/e)$ and $f(G/e) : M(G/e) \longrightarrow N(G/e)$ are both monomorphisms. □

The remainder of this chapter is devoted to the proof of Theorem 8.1. For $p \neq 2$, our argument is essentially that given in [**6**]. For $p = 2$, it is an extension of the computation of H_* given in [**12**]. Throughout our proof, ξ denotes a nontrivial irreducible complex representation of G, and ζ denotes the real one-dimensional sign representation of $\mathbb{Z}/2$. The key tools for our computation of H_*^S are the equivariant cofibre sequences

$$(G/e)_+ \xrightarrow{\iota_\xi} S\xi_+ \xrightarrow{\pi'_\xi} \Sigma(G/e)_+,$$

$$S^0 \xrightarrow{\epsilon_\xi} S^\xi \xrightarrow{\pi_\xi} \Sigma S\xi_+,$$

and

$$S^0 \xrightarrow{\epsilon_\zeta} S^\zeta \xrightarrow{\pi_\zeta} \Sigma(G/e)_+,$$

the last of which applies only for $p = 2$.

REMARK 8.4. Several observations about these cofibre sequences are needed in our computations.

(a) The next map on the right in our first cofibre sequence is of the form

$$\Sigma(G/e)_+ \xrightarrow{1-g} \Sigma(G/e)_+$$

for some generator g of G.

(b) The next map on the right in our second cofibre sequence is the suspension of the collapse map $\phi : S\xi_+ \longrightarrow S^0$ which sends all of $S\xi$ to the non-basepoint of S^0.

(c) The composite

$$(G/e)_+ \xrightarrow{\iota_\xi} S\xi_+ \xrightarrow{\phi} S^0$$

is just the geometric restriction map $(G/e)_+ \longrightarrow S^0$ which collapses all of G/e to the non-basepoint of S^0. In cohomology, this map induces the restriction map. In homology it induces the transfer since its Spanier-Whitehead dual is the geometric transfer map (see Corollary III.5.2 of [**18**]).

(d) The composite

$$S^\xi \xrightarrow{\pi_\xi} \Sigma S\xi_+ \xrightarrow{\Sigma\pi'_\xi} \Sigma^2(G/e)_+$$

is related to the geometric transfer map $\tau_\xi : S^\xi \longrightarrow \Sigma^\xi(G/e)_+$ by the commuting diagram

$$\begin{array}{ccc} S^\xi & \xrightarrow{\pi_\xi} & \Sigma S\xi_+ \\ {\scriptstyle\tau_\xi}\downarrow & & \downarrow{\scriptstyle\Sigma\pi'_\xi} \\ \Sigma^\xi(G/e)_+ & \xrightarrow{\lambda} & \Sigma^2(G/e)_+. \end{array}$$

Here, λ is a special case of the G-homeomorphism $\lambda : (G/e)_+ \wedge X \longrightarrow (G/e)_+ \wedge X$, available for any G-space X, which relates $(G/e)_+ \wedge X$ with G acting diagonally to $(G/e)_+ \wedge X$ with G acting only on $(G/e)_+$. This map is given by $\lambda(g, x) = (g, g^{-1}x)$.

(e) The commuting diagram

$$\begin{array}{ccc} S^\zeta & \xrightarrow{\tau_\zeta} & \Sigma^\zeta(G/e)_+ \\ & {\scriptstyle\pi_\zeta}\searrow & \downarrow{\scriptstyle\lambda} \\ & & \Sigma(G/e)_+ \end{array}$$

identifies the map π_ζ of our third cofibre sequence with the geometric transfer map $\tau_\zeta : S^\zeta \longrightarrow \Sigma^\zeta(G/e)_+$.

Our next two results generalize Lemma A.1 of [**12**].

PROPOSITION 8.5. *Let ω be an element of $RO(G)$. Then*

(a)

$$\widetilde{H}^G_\omega(S\xi_+;S) = \begin{cases} \mathfrak{L}(H^S_\omega) & \textit{if } |\omega| = 0, \\ \mathfrak{R}(H^S_{\omega-1}) & \textit{if } |\omega| = 1, \\ 0 & \textit{otherwise.} \end{cases}$$

Moreover, if $|\omega| = 0$, *then the diagram*

$$\begin{array}{ccc} \mathfrak{L}(H^S_\omega) & \xrightarrow{\cong} & \widetilde{H}^G_\omega(S\xi_+;S) \\ & \searrow^{\bar{\epsilon}_{H^S_\omega}} & \downarrow{\scriptstyle \phi_*} \\ & & H^S_\omega \end{array}$$

commutes.

(b)

$$\widetilde{H}_G^\omega(S\xi_+;S) = \begin{cases} \mathfrak{R}(\widetilde{H}^\omega_G(S^0;S)) \cong \mathfrak{R}(H^S_{-\omega}) & \textit{if } |\omega| = 0, \\ \mathfrak{L}(\widetilde{H}^{\omega-1}_G(S^0;S)) \cong \mathfrak{L}(H^S_{1-\omega}) & \textit{if } |\omega| = 1, \\ 0 & \textit{otherwise.} \end{cases}$$

Moreover, if $|\omega| = 0$, *then the diagram*

$$\begin{array}{ccc} \widetilde{H}^\omega_G(S^0;S) & & \\ \downarrow{\scriptstyle \phi^*} & \searrow^{\bar{\eta}} & \\ \widetilde{H}^\omega_G(S\xi_+;S) & \xrightarrow{\cong} & \mathfrak{R}(\widetilde{H}^\omega_G(S^0;S)) \end{array}$$

commutes.

PROOF. Geometrically, the key to this proof is the observation in Remark 8.4(a) about the next map on the right in our first cofibre sequence. Algebraically, the key is the fact that, since G is cyclic, the Mackey functors $\mathfrak{L}(M)$ and $\mathfrak{R}(M)$ and the canonical maps $\epsilon_M : \mathfrak{L}(M) \longrightarrow M$ and $\eta_M : M \longrightarrow \mathfrak{R}(M)$ are specified by the commuting diagram

$$\begin{array}{ccccccccccc} & & M & & & & & & & & \\ & & \downarrow{\scriptstyle \eta_M} & \searrow^{\widehat{\rho}_M} & & & & & & & \\ 0 & \longrightarrow & \mathfrak{R}(M) & \longrightarrow & M_{G/e} & \xrightarrow{1-g} & M_{G/e} & \longrightarrow & \mathfrak{L}(M) & \longrightarrow & 0 \\ & & & & & & & \searrow_{\widehat{\tau}_M} & \downarrow{\scriptstyle \epsilon_M} & & \\ & & & & & & & & M & & \end{array} \tag{8.1}$$

in which the long row is exact. The middle map in this row is the difference of the self maps of $M_{G/e}$ induced by the identity map and the multiplication by g map on G/e.

The homology and cohomology long exact sequences associated to our first cofibre sequence can be compared to the row in this diagram via the isomorphisms $\widetilde{H}^G_\omega((G/e)_+;S) \cong (H^S_\omega)_{G/e}$ and $\widetilde{H}^\omega_G((G/e)_+;S) \cong (\widetilde{H}^\omega_G(S^0;S))_{G/e}$. This comparison yields the short exact sequences

$$0 \longrightarrow \mathfrak{L}(H^S_\omega) \longrightarrow \widetilde{H}^G_\omega(S\xi_+;S) \longrightarrow \mathfrak{R}(H^S_{\omega-1}) \longrightarrow 0$$

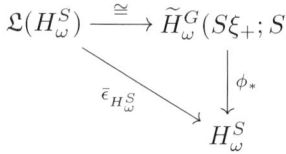

and

$$0 \longrightarrow \mathfrak{L}(H^S_{1-\omega}) \longrightarrow \widetilde{H}^\omega_G(S\xi_+; S) \longrightarrow \mathfrak{R}(H^S_{-\omega}) \longrightarrow 0.$$

For any Mackey functor M, $\mathfrak{L}(M)$ and $\mathfrak{R}(M)$ are completely determined by the G-module $M(G/e)$. Since H^S_ω vanishes at G/e unless $|\omega| = 0$, it follows immediately that $\widetilde{H}^G_\omega(S\xi_+; S)$ and $\widetilde{H}^\omega_G(S\xi_+; S)$ vanish unless $|\omega|$ is 0 or 1. If $|\omega|$ is 0 or 1, the asserted values of $\widetilde{H}^G_\omega(S\xi_+; S)$ and $\widetilde{H}^\omega_G(S\xi_+; S)$ follow immediately from the short exact sequences. The commutativity of the two diagrams follows directly from the characterization of the maps $\widehat{\rho}_M$ and $\widehat{\tau}_M$ given by diagram (8.1). $\square$

Explicit values for $\widetilde{H}^G_\omega(S\xi_+; S)$ and $\widetilde{H}^\omega_G(S\xi_+; S)$ can be obtained by looking more closely at $H^S_\omega(G/e)$.

COROLLARY 8.6. *For* $\omega \in RO(G)$,

$$\widetilde{H}^G_\omega(S\xi_+; S) = \begin{cases} \mathfrak{L}(S) & \text{if } |\omega| = 0 \text{ and } |\omega^G| \text{ is even,} \\ \mathfrak{L}_-(S) & \text{if } |\omega| = 0 \text{ and } |\omega^G| \text{ is odd,} \\ \mathfrak{R}(S) & \text{if } |\omega| = 1 \text{ and } |\omega^G| \text{ is odd,} \\ \mathfrak{R}_-(S) & \text{if } |\omega| = 1 \text{ and } |\omega^G| \text{ is even,} \\ 0 & \text{otherwise.} \end{cases}$$

and

$$\widetilde{H}^\omega_G(S\xi_+; S) = \begin{cases} \mathfrak{R}(S) & \text{if } |\omega| = 0 \text{ and } |\omega^G| \text{ is even,} \\ \mathfrak{R}_-(S) & \text{if } |\omega| = 0 \text{ and } |\omega^G| \text{ is odd,} \\ \mathfrak{L}(S) & \text{if } |\omega| = 1 \text{ and } |\omega^G| \text{ is odd,} \\ \mathfrak{L}_-(S) & \text{if } |\omega| = 1 \text{ and } |\omega^G| \text{ is even,} \\ 0 & \text{otherwise.} \end{cases}$$

PROOF. If $|\omega| = 0$, then the abelian group $H^S_\omega(G/e)$ is just $S(G/e)$. However, the actions of G on $H^S_\omega(G/e)$ and $S(G/e)$ need not be the same due to the action of G on the sphere S^ω. If $|\omega^G|$ is even, then the two G-actions agree since the action of G on S^ω is homologically trivial. However, if $|\omega^G|$ is odd, then $p = 2$ and ω contains an odd number of copies of the real sign representation ζ. In this case, $H^S_\omega(G/e)$ is isomorphic as a G-module to $S(G/e) \otimes \mathbb{Z}_-$. This description of the G-action on $H^S_\omega(G/e)$, together with the descriptions of $\widetilde{H}^G_\omega(S\xi_+; S)$ and $\widetilde{H}^\omega_G(S\xi_+; S)$ given in Proposition 8.5, suffices to complete the proof. $\square$

The proposition above, coupled with our second and third cofibre sequences, yields the following variant of Lemma A.2 of [**12**]. Here, and elsewhere, the homology and cohomology maps induced by the maps $\epsilon_\xi : S^0 \longrightarrow S^\xi$ and $\epsilon_\zeta : S^0 \longrightarrow S^\zeta$ are also denoted by ϵ_ξ and ϵ_ζ.

LEMMA 8.7. *Let* ω *be an element of* $RO(G)$.

(a) *The map* $\epsilon_\xi : H^S_{\omega+\xi} \longrightarrow H^S_\omega$ *is*

$$\begin{cases} \textit{epi} & \text{if } |\omega| \neq 0, -1, \\ \textit{mono} & \text{if } |\omega| \neq -1, -2, \\ \textit{iso} & \text{if } |\omega| \neq 0, -1, -2. \end{cases}$$

(b) *If $p = 2$, then the map $\epsilon_\zeta : H^S_{\omega+\zeta} \longrightarrow H^S_\omega$ is*

$$\begin{cases} \textit{epi} & \textit{if } |\omega| \neq 0, \\ \textit{mono} & \textit{if } |\omega| \neq -1, \\ \textit{iso} & \textit{if } |\omega| \neq 0, -1. \end{cases}$$

(c) *If $S(G/e) = 0$, then the map $\epsilon_\xi : H^S_{\omega+\xi} \longrightarrow H^S_\omega$ and, for $p = 2$, the map $\epsilon_\zeta : H^S_{\omega+\zeta} \longrightarrow H^S_\omega$ are isomorphisms for all $\omega \in RO(G)$.*

This result plus the dimension axiom gives the vanishing of H^S_ω for any ω which plots in either the first or third quadrant, but not on the coordinate axis (see also Lemma A.3 of [**12**]).

LEMMA 8.8. *Let ω be an element of $RO(G)$. If $|\omega|$ and $|\omega^G|$ are either both positive or both negative, then $H^S_\omega = 0$.*

PROOF. Consider the special case in which $\omega = n+\xi_1-\xi_2$, where n is a positive integer and ξ_1, ξ_2 are irreducible complex representations of G. Then $H^S_n = 0$ by the dimension axiom. The pair of isomorphisms

$$H^S_{n+\xi_1-\xi_2} \xleftarrow[\cong]{\epsilon_{\xi_2}} H^S_{n+\xi_1} \xrightarrow[\cong]{\epsilon_{\xi_1}} H^S_n = 0$$

then gives the vanishing of H^S_ω. The general case follows by the obvious extension of this argument. □

Lemma 8.7 indicates that all of H^S_* can be determined from its values at those $\omega \in RO(G)$ such that $-2 \leq |\omega| \leq 2$. The following result, which generalizes Lemmas A.4, A.8, and A.9 of [**12**], further reduces the computation of H^S_* down to that of its values at those $\omega \in RO(G)$ such that $|\omega| = 0$.

PROPOSITION 8.9. *Let ω be an element of $RO(G)$.*

(a) *If $|\omega| = -2$, then*

$$H^S_\omega = \text{Coker}\left(\widehat{\tau} : (H^S_{\omega+\xi})_{G/e} \longrightarrow H^S_{\omega+\xi}\right).$$

(b) *If $|\omega| = 2$, then*

$$H^S_\omega = \text{Ker}\left(\widehat{\rho} : H^S_{\omega-\xi} \longrightarrow (H^S_{\omega-\xi})_{G/e}\right).$$

(c) *If $|\omega| = -1$ and $|\omega^G| > 0$, then*

$$H^S_\omega = \text{Ker}\left(\epsilon_{H^S_{\omega+\xi-1}} : \mathfrak{L}(H^S_{\omega+\xi-1}) \longrightarrow H^S_{\omega+\xi-1}\right).$$

(d) *If $|\omega| = 1$ and $|\omega^G| < 0$, then*

$$H^S_\omega = \text{Coker}\left(\bar{\eta}_{H^S_{\omega-\xi+1}} : H^S_{\omega-\xi+1} \longrightarrow \mathfrak{R}(H^S_{\omega-\xi+1})\right).$$

(e) *If $p = 2$ and $|\omega| = -1$, then*

$$H^S_\omega = \text{Coker}\left(\widehat{\tau} : (H^S_{\omega+\zeta})_{G/e} \longrightarrow H^S_{\omega+\zeta}\right).$$

(f) *If $p = 2$ and $|\omega| = 1$, then*

$$H^S_\omega = \text{Ker}\left(\widehat{\rho} : H^S_{\omega-\zeta} \longrightarrow (H^S_{\omega-\zeta})_{G/e}\right).$$

PROOF. For part (a), assume $|\omega| = -2$ and consider the diagram

$$\begin{array}{ccccccc}
\widetilde{H}^G_{\omega+\xi}((G/e)_+;S) & & & & & & \\
\downarrow{\scriptstyle (\iota_\xi)_*} & \searrow^{\widehat{\tau}} & & & & & \\
\widetilde{H}^G_{\omega+\xi}(S\xi_+;S) & \xrightarrow{\phi_*} & \widetilde{H}^G_{\omega+\xi}(S^0;S) & \longrightarrow & \widetilde{H}^G_{\omega+\xi}(S^\xi;S) & \longrightarrow & \widetilde{H}^G_{\omega+\xi}(\Sigma S\xi_+;S) = 0
\end{array}$$

in which the exact row comes from our second cofibre sequence. The Mackey functor on the right vanishes by Proposition 8.5 since $|\omega + \xi - 1| = -1$. Thus, $H^S_\omega \cong \widetilde{H}^G_{\omega+\xi}(S^\xi;S)$ is the cokernel of ϕ_*. However, $(\iota_\xi)_*$ is the surjective map displaying $\widetilde{H}^G_{\omega+\xi}(S\xi_+;S)$ as the quotient $\mathfrak{L}(H^S_{\omega+\xi})$ of $\widetilde{H}^G_{\omega+\xi}((G/e)_+;S)$. Thus, H^S_ω is also the cokernel of $\widehat{\tau}$.

For part (b), assume $|\omega| = 2$ and consider the diagram

$$\begin{array}{ccccccc}
0 = \widetilde{H}_G^{\xi-\omega}(\Sigma S\xi_+;S) & \longrightarrow & \widetilde{H}_G^{\xi-\omega}(S^\xi;S) & \longrightarrow & \widetilde{H}_G^{\xi-\omega}(S^0;S) & \xrightarrow{\phi^*} & \widetilde{H}_G^{\xi-\omega}(S\xi_+;S) \\
 & & & & & \searrow^{\widehat{\rho}} & \downarrow{\scriptstyle (\iota_\xi)^*} \\
 & & & & & & \widetilde{H}_G^{\xi-\omega}((G/e)_+;S)
\end{array}$$

in which the exact row comes from our second cofibre sequence. Here, the left-most Mackey functor vanishes by Proposition 8.5 so that $H^S_\omega \cong \widetilde{H}_G^{\xi-\omega}(S^\xi;S)$ is the kernel of ϕ^*. In this case, $(\iota_\xi)^*$ is the injective map displaying $\widetilde{H}_G^{\xi-\omega}(S\xi_+;S)$ as the subobject $\mathfrak{R}(H^S_{\omega-\xi})$ of $\widetilde{H}_G^{\xi-\omega}((G/e)_+;S)$. Thus, H^S_ω is the kernel of $\widehat{\rho}$.

For part (c), assume $|\omega| = -1$ and $|\omega^G| > 0$, and consider the exact sequence

$$0 = \widetilde{H}^G_{\omega+\xi}(S^0;S) \longrightarrow \widetilde{H}^G_{\omega+\xi}(S^\xi;S) \longrightarrow \widetilde{H}^G_{\omega+\xi}(\Sigma S\xi_+;S) \xrightarrow{(\Sigma\phi)_*} \widetilde{H}^G_{\omega+\xi}(S^1;S)$$

derived from our second cofibre sequence. Here, the left-most Mackey functor vanishes by Lemma 8.8. Thus, $H^S_\omega \cong \widetilde{H}^G_{\omega+\xi}(S^\xi;S)$ is the kernel of $(\Sigma\phi)_*$. Proposition 8.5(a) provides the identification of $(\Sigma\phi)_*$ with $\bar{\epsilon}_{H^S_{\omega+\xi-1}} : \mathfrak{L}(H^S_{\omega+\xi-1}) \longrightarrow H^S_{\omega+\xi-1}$.

Part (d) is handled in much the same way as part (c) by examining the exact sequence

$$\widetilde{H}_G^{\xi-\omega}(S^1;S) \xrightarrow{(\Sigma\phi)^*} \widetilde{H}_G^{\xi-\omega}(\Sigma S\xi_+;S) \longrightarrow \widetilde{H}_G^{\xi-\omega}(S^\xi;S) \longrightarrow \widetilde{H}_G^{\xi-\omega}(S^0;S) = 0$$

and using Proposition 8.5(b) to identify the map $(\Sigma\phi)^*$.

Now assume that $p = 2$ and that $|\omega| = -1$. Part (e) follows immediately from the exact sequence

$$\widetilde{H}_G^{-\omega}(\Sigma^\zeta(G/e)_+;S) \xrightarrow{\widehat{\tau}} \widetilde{H}_G^{-\omega}(S^\zeta;S) \longrightarrow \widetilde{H}_G^{-\omega}(S^0;S) \longrightarrow \widetilde{H}_G^{-\omega}((G/e)_+;S) = 0$$

derived from our third cofibre sequence. Notice that we have used the identification of the map π_ζ in that sequence with the transfer. Part (f) follows similarly from the exact sequence

$$0 = \widetilde{H}_G^{\zeta-\omega}(\Sigma(G/e)_+;S) \longrightarrow \widetilde{H}_G^{\zeta-\omega}(S^\zeta;S) \longrightarrow \widetilde{H}_G^{\zeta-\omega}(S^0;S) \xrightarrow{\widehat{\rho}} \widetilde{H}_G^{\zeta-\omega}((G/e)_+;S)$$

and the isomorphism $H^S_\omega \cong \widetilde{H}_G^{\zeta-\omega}(S^\zeta;S)$. $\square$

REMARK 8.10. If $p = 2$, then any one of the three pairs (a) and (b), (c) and (d), or (e) and (f) of parts of this proposition can be used to derive all the values of H_ω^S from those for which $|\omega| = 0$. However, if p is odd, then all four of parts (a), (b), (c) and (d) are needed since, for any $\omega \in RO(G)$, $|\omega|$ and $|\omega^G|$ are either both even or both odd.

Our next two results, which generalize Lemmas A.5 and A.6 of [**12**], give the values of H_ω^S for those $\omega \in RO(G)$ such that $|\omega| = 0$.

PROPOSITION 8.11. *Let ω be an element of $RO(G)$ such that $|\omega| = 0$.*

(a) *If $|\omega^G| < -1$, then $H_\omega^S \cong \widetilde{H}_\omega^G(S\xi_+; S)$.*
(b) *If $|\omega^G| > 1$, then $H_\omega^S \cong \widetilde{H}_G^{-\omega}(S\xi_+; S)$.*
(c) *If $p = 2$ and $|\omega^G| = -1$, then $H_\omega^S \cong \mathcal{C}(S)$.*
(d) *If $p = 2$ and $|\omega^G| = 1$, then $H_\omega^S \cong \mathcal{K}(S)$.*

PROOF. For part (a), consider the exact sequence

$$\widetilde{H}_\omega^G(S^{\xi-1}; S) \longrightarrow \widetilde{H}_\omega^G(S\xi_+; S) \longrightarrow \widetilde{H}_\omega^G(S^0; S) \longrightarrow \widetilde{H}_\omega^G(S^\xi; S)$$

derived from our second cofibre sequence. The first and last terms in this sequence are isomorphic to $H_{\omega-\xi+1}^S$ and $H_{\omega-\xi}^S$, respectively. These vanish by Lemma 8.8.

For part (b), consider the analogous cohomology exact sequence

$$\widetilde{H}_G^{-\omega}(S^\xi; S) \longrightarrow \widetilde{H}_G^{-\omega}(S^0; S) \longrightarrow \widetilde{H}_G^{-\omega}(S\xi_+; S) \longrightarrow \widetilde{H}_G^{-\omega}(S^{\xi-1}; S).$$

Here, the first and last terms are isomorphic to $H_{\omega+\xi}^S$ and $H_{\omega+\xi-1}^S$, respectively. These also vanish by Lemma 8.8.

In part (c), ω must be $\zeta - 1$. Our third cofibre sequence yields the cohomology exact sequence

$$\widetilde{H}_G^0(S^0; S) \longrightarrow \widetilde{H}_G^0((G/e)_+; S) \longrightarrow \widetilde{H}_G^1(S^\zeta; S) \longrightarrow \widetilde{H}_G^1(S^0; S).$$

By the dimension axiom, the last term in this sequence is zero and the first two are S and $S_{G/e}$, respectively. The third term, which is isomorphic to $H_{\zeta-1}^S$, must therefore also be isomorphic to $\mathcal{C}(S)$.

In part (d), ω must be $1 - \zeta$. Our third cofibre sequence yields the homology exact sequence

$$\widetilde{H}_1^G(S^0; S) \longrightarrow \widetilde{H}_1^G(S^\zeta; S) \longrightarrow \widetilde{H}_0^G((G/e)_+; S) \longrightarrow \widetilde{H}_0^G(S^0; S).$$

By the dimension axiom, the first term in this sequence is zero and the last two are $S_{G/e}$ and S, respectively. The second term, which is isomorphic to $H_{1-\zeta}^S$, must therefore also be isomorphic to $\mathcal{K}(S)$. □

PROPOSITION 8.12. *Let X be a G-space, $v \in RO(G)$, and $\omega \in RO_0(G)$. Then the product map*

$$H_\omega \,\square\, \widetilde{H}_v^G(X; S) \longrightarrow \widetilde{H}_{\omega+v}^G(X; S)$$

is an isomorphism. In particular, $\widetilde{H}_\omega^G(X; S) \cong H_\omega \,\square\, \widetilde{H}_0^G(X; S)$.

PROOF. Apply Theorem 3.2 of [**5**] to the G-spectrum $S^{-\omega}$, which is obviously invertible. Our assumption that $|\omega| = |\omega^G| = 0$ is equivalent to the assertion that the function d of that theorem (which is not our function d) vanishes on $S^{-\omega}$.

Thus, by that theorem, $S^{-\omega}$ is a Künneth object. Proposition 1.2 of [**5**] therefore indicates that $S^{-\omega}$ is a retract of a wedge of copies of S^0.

The multiplication map

$$\widetilde{H}_0^G(S^0; A) \,\Box\, \widetilde{H}_v^G(X; S) \longrightarrow \widetilde{H}_v^G(S^0 \wedge X; S) \cong \widetilde{H}_v^G(X; S)$$

is just the unit isomorphism for the action of H_* on $\widetilde{H}_*^G(X; S)$. It follows easily that the map

$$\widetilde{H}_0^G(S^0; A) \,\Box\, \widetilde{H}_v^G(X; S) \longrightarrow \widetilde{H}_v^G(S^0 \wedge X; S)$$

remains an isomorphism if S^0 is replaced by either a wedge of copies of S^0 or a retract of a wedge of copies of S^0. The obvious identification of $\widetilde{H}_0^G(S^{-\omega}; A)$ with H_ω suffices to complete the proof. $\Box$

In the proof of our main freeness theorem, we need to know that an analog of this proposition holds for any module over H_* rather than just those modules arising in homology.

COROLLARY 8.13. *Let N be a module over H_*, $v \in RO(G)$, and $\omega \in RO_0(G)$. Then the action map*

$$\nu_{\omega,v} : H_\omega \,\Box\, N_v \longrightarrow N_{\omega+v}$$

is an isomorphism.

PROOF. In the commuting diagram

$$\begin{array}{ccc} H_\omega \,\Box\, H_{-\omega} \,\Box\, N_{\omega+v} & \xrightarrow[\cong]{\mu\Box 1} & H_0 \,\Box\, N_{\omega+v} \\ {\scriptstyle 1\Box\nu_{-\omega,\omega+v}}\downarrow & & \downarrow{\scriptstyle \cong\ \nu_{0,\omega+v}} \\ H_\omega \,\Box\, N_v & \xrightarrow[\nu_{\omega,v}]{} & N_{\omega+v} \end{array}$$

the top horizontal map is an isomorphism by the proposition, and the right vertical map is the unit isomorphism for the action of H_* on N. Thus, the map $1\Box\nu_{-\omega,\omega+v}$ is a monomorphism. Similarly, the top horizontal and right vertical maps in the commuting diagram

$$\begin{array}{ccc} H_{-\omega} \,\Box\, H_\omega \,\Box\, N_v & \xrightarrow[\cong]{\mu\Box 1} & H_0 \,\Box\, N_v \\ {\scriptstyle 1\Box\nu_{\omega,v}}\downarrow & & \downarrow{\scriptstyle \cong\ \nu_{0,v}} \\ H_{-\omega} \,\Box\, N_{\omega+v} & \xrightarrow[\nu_{-\omega,\omega+v}]{} & N_v \end{array}$$

are isomorphisms. Thus, the map $\nu_{-\omega,\omega+v}$ is an epimorphism. But then the right exactness of box products implies that the map

$$1 \,\Box\, \nu_{-\omega,\omega+v} : H_\omega \,\Box\, H_{-\omega} \,\Box\, N_{\omega+v} \longrightarrow H_\omega \,\Box\, N_v$$

must also be an epimorphism. It is therefore an isomorphism. The first commuting square then gives that $\nu_{\omega,v}$ is an isomorphism. $\Box$

REMARK 8.14. To use Proposition 8.12 in the proof of Theorem 8.1, we need to identify H_ω with $A[d_\omega]$. This is done in detail in Lemma A.12 of [**12**]. However, a quick summary of that argument is easily given. The portions

$$\cdots \longrightarrow \widetilde{H}_\omega^G(S\xi_+; A) \longrightarrow \widetilde{H}_\omega^G(S^0; A) \xrightarrow{\epsilon_\xi} \widetilde{H}_\omega^G(S^\xi; A) \longrightarrow \cdots$$

and

$$\cdots \longrightarrow \widetilde{H}_G^{-\omega}(S^\xi; A) \longrightarrow \widetilde{H}_G^{-\omega}(S^0; A) \longrightarrow \widetilde{H}_G^{-\omega}(S\xi_+; A) \longrightarrow \cdots$$

of the homology and cohomology long exact sequences coming from our second cofibre sequence reduce to the short exact sequences

$$0 \longrightarrow L \longrightarrow H_\omega \xrightarrow{\epsilon_\xi} \langle\mathbb{Z}\rangle \longrightarrow 0$$

and

$$0 \longrightarrow \langle\mathbb{Z}\rangle \longrightarrow H_\omega \longrightarrow R \longrightarrow 0,$$

respectively. By Lemma 12.2(c), the only common solutions to these two extension problems are of the form $A[d]$, for some integer d prime to p. The appropriate d is determined by the fact that there is an element of $A[d](G/G)$ which restricts to d in $A[d](G/e)$ and maps under ϵ_ξ to a generator of $\langle\mathbb{Z}\rangle(G/G)$ in the first short exact sequence above.

Now consider the special case in which $\omega = \xi - \xi'$, where ξ' is another irreducible complex G-representation. Both ξ and ξ' are copies of the complex plane on which G acts by multiplication by p^{th} roots of unity. Thus, there is a integer m prime to p such that the complex power map taking z to z^m is an equivariant map from ξ to ξ'. This map extends to an equivariant map f from S^ξ to $S^{\xi'}$, which may be regarded as an element of $H_\omega(G/G)$ via the Hurewicz map. It is easy to see that $\epsilon_\xi(f)$ is a generator of $\langle\mathbb{Z}\rangle$. Clearly the restriction of f to $H_\omega(G/e)$ is just its degree m, which is d_ω by our definition of d_ω. Since the element τ of $A[d](G/G)$ restricts to p in $A[d](G/e)$ and maps to zero under ϵ_ξ, this argument only determines d modulo p. Moreover, since we haven't worried about the orientation of $\langle\mathbb{Z}\rangle$, we have really determined d only up to sign. However, $A[d] \cong A[d']$ if and only if $d \equiv \pm d' \mod p$, so this level of imprecision is irrelevant. The argument for a general element ω of $RO_0(G)$ is just the obvious extension of this argument.

The following consequence of Corollary 8.13 is not used in our computation of H_*^S, but is needed for the proof of our main freeness theorem.

COROLLARY 8.15. *Let N be a module over H_*. If there is an element υ of $RO(G)$ and an integer d prime to p such that $N_\upsilon \cong A[d]$, then there is an element υ' of $RO(G)$ such that $N_{\upsilon'} \cong A$, $|\upsilon'| = |\upsilon|$, and $|(\upsilon')^G| = |\upsilon^G|$.*

PROOF. Select an element ω of $RO_0(G)$ such that $d_\omega d \equiv 1 \bmod p$. The map

$$\nu_{\omega,\upsilon} : H_\omega \,\Box\, N_\upsilon \longrightarrow N_{\omega+\upsilon}$$

is an isomorphism by Corollary 8.13. By Remark 8.14, $H_\omega \cong A[d_\omega]$. By Table 1.1, $A[d_\omega] \,\Box\, A[d] \cong A[d_\omega d]$; and, by Lemma 1.1, $A[d_\omega d] \cong A[1] = A$. Thus, $\upsilon' = \omega + \upsilon$ is the desired element of $RO(G)$. □

Completing the proof of this chapter's main theorem requires nothing more than tying the preceding results together properly.

PROOF OF THEOREM 8.1. First note that parts (ii) and (iii) follow immediately from Proposition 8.11 and Corollary 8.6. Also note that part (iv) is just a restatement of Lemma 8.8. Similarly, the middle of the three values described in part (i) comes directly from Proposition 8.12 and Remark 8.14.

To verify the other two values given in part (i), first observe that

$$H^S_{-\xi} = \text{Coker}\left(\widehat{\tau}_S : S_{G/e} \longrightarrow S\right)$$

and

$$H^S_{\xi} = \text{Ker}\left(\widehat{\rho}_S : S \longrightarrow S_{G/e}\right)$$

by parts (a) and (b) of Proposition 8.9. If $p = 2$, then

$$H^S_{-\zeta} = \text{Coker}\left(\widehat{\tau}_S : S_{G/e} \longrightarrow S\right)$$

and

$$H^S_{\zeta} = \text{Ker}\left(\widehat{\rho}_S : S \longrightarrow S_{G/e}\right)$$

by parts (e) and (f) of the same proposition. For $p = 2$, Lemma 8.7 obviously suffices to complete the proof since the only elements ω of $RO(G)$ satisfying $|\omega^G| = 0$ and $|\omega| \neq 0$ are of the form $m\zeta$ for some integer m. If $p \neq 2$, such an ω is a linear combination $\xi_1 + \xi_2 + \ldots + \xi_m - \eta_1 - \eta_2 - \ldots - \eta_n$ of nontrivial irreducible complex G-representations in which $m \neq n$. For the special case in which $m = 2$ and $n = 1$, the pair of isomorphisms

$$H^S_{\xi_1+\xi_2-\eta_1} \xleftarrow[\cong]{\epsilon_{\eta_1}} H^S_{\xi_1+\xi_2} \xrightarrow[\cong]{\epsilon_{\xi_2}} H^S_{\xi_1}$$

suffices to complete the proof. Longer chains of isomorphisms of the same sort suffice to handle any case in which $m > n$. A very similar chain of isomorphisms argument handles the cases for which $m < n$.

For part (v), observe that Proposition 8.9(a) and part (ii) of Theorem 8.1 indicate that the first value should be $\text{Coker}\left(\widehat{\tau}_{\mathfrak{R}(S)} : \mathfrak{R}(S)_{G/e} \longrightarrow \mathfrak{R}(S)\right)$. The naturality of the maps $\widehat{\tau}$ implies that the diagram

$$\begin{array}{ccc}
\mathfrak{L}(S)_{G/e} & \xrightarrow{\widehat{\tau}_{\mathfrak{L}(S)}} & \mathfrak{L}(S) \\
{\scriptstyle (\bar{\eta}_S \circ \bar{\epsilon}_S)_{G/e}} \downarrow & & \downarrow {\scriptstyle \bar{\eta}_S \circ \bar{\epsilon}_S} \\
\mathfrak{R}(S)_{G/e} & \xrightarrow{\widehat{\tau}_{\mathfrak{R}(S)}} & \mathfrak{R}(S)
\end{array}$$

commutes. The left vertical map in this diagram is an isomorphism since $\bar{\eta}_S \circ \bar{\epsilon}_S$ is an isomorphism at G/e. The top horizontal map in the diagram is an epimorphism since the transfer map of $\mathfrak{L}(S)$ is an epimorphism. Thus, the maps $\widehat{\tau}_{\mathfrak{R}(S)}$ and $\bar{\eta}_S \circ \bar{\epsilon}_S$ have isomorphic cokernels. The second value in part should (v) be $\text{Ker}\left(\bar{\epsilon}_{\mathfrak{R}(S)} : \mathfrak{L}(\mathfrak{R}(S)) \longrightarrow \mathfrak{R}(S)\right)$ by Proposition 8.9(c) and part (ii) of this theorem. Note, however, that $\mathfrak{L}(\mathfrak{R}(S)) \cong \mathfrak{L}(S)$ since $\mathfrak{R}(S)(G/e)$ and $S(G/e)$ are equal. Moreover, under this isomorphism, the map $\bar{\epsilon}_{\mathfrak{R}(S)}$ is identified with the composite $\bar{\eta}_S \circ \bar{\epsilon}_S$ since each of these maps is uniquely determined by the fact that it is the identity map at G/e.

For the third value in part (v), first consider the special case in which $\omega = 1 - \xi$. Then, from Proposition 8.9(c), we have that

$$H^S_{1-\xi} = \text{Ker}\left(\bar{\epsilon}_S : \mathfrak{L}(S) \longrightarrow S\right).$$

If $p = 2$, then Lemma 8.7 suffices to complete the proof since the only elements ω satisfying the appropriate conditions are of the form $1 - m\zeta$, where $m \geq 2$. Here, we use the fact that $\xi = 2\zeta$ in $RO(G)$. For $p \neq 2$, ω must be of the form

$1+\eta_1+\eta_2+\ldots+\eta_m-\xi_1-\xi_2-\ldots-\xi_n$, with $0\leq m<n$. For the special case in which $m=1$ and $n=2$, the pair of isomorphisms

$$H^S_{1+\eta_1-\xi_1-\xi_2} \xrightarrow[\cong]{\epsilon_{\eta_1}} H^S_{1-\xi_1-\xi_2} \xleftarrow[\cong]{\epsilon_{\xi_2}} H^S_{1-\xi_1}$$

completes the proof. Longer chains of isomorphisms of the same sort suffice to handle the general case.

For part (vi), note that Proposition 8.9(b) and part (iii) of this theorem give that the first value should be Ker $\left(\widehat{\rho}_{\mathfrak{L}(S)}:\mathfrak{L}(S)\longrightarrow\mathfrak{L}(S)_{G/e}\right)$. The naturality of the maps $\widehat{\rho}$ indicates that the diagram

$$\begin{array}{ccc}\mathfrak{L}(S) & \xrightarrow{\widehat{\rho}_{\mathfrak{L}(S)}} & \mathfrak{L}(S)_{G/e}\\ {\scriptstyle \bar{\eta}_S\circ\bar{\epsilon}_S}\downarrow & & \downarrow{\scriptstyle (\bar{\eta}_S\circ\bar{\epsilon}_S)_{G/e}}\\ \mathfrak{R}(S) & \xrightarrow{\widehat{\rho}_{\mathfrak{R}(S)}} & \mathfrak{R}(S)_{G/e}\end{array}$$

commutes. The right vertical map in this diagram is an isomorphism since $\bar{\eta}_S\circ\bar{\epsilon}_S$ is an isomorphism at G/e. Moreover, the bottom horizontal map is a monomorphism since the restriction map of $\mathfrak{R}(S)$ is a monomorphism. Thus, the maps $\widehat{\rho}_{\mathfrak{L}(S)}$ and $\bar{\eta}_S\circ\bar{\epsilon}_S$ have isomorphic kernels. Proposition 8.9(d) and part (ii) of this theorem indicate that the second value should be Coker $\left(\bar{\eta}_{\mathfrak{L}(S)}:\mathfrak{L}(S)\longrightarrow\mathfrak{R}(\mathfrak{L}(S))\right)$. However, $\mathfrak{R}(\mathfrak{L}(S))$ and $\mathfrak{R}(S)$ are isomorphic since $\mathfrak{L}(S)(G/e)$ and $S(G/e)$ are equal. As in part (v), this isomorphism identifies the map $\bar{\eta}_{\mathfrak{L}(S)}$ with the composite $\bar{\eta}_S\circ\bar{\epsilon}_S$. For the third value, look first at the special case in which $\omega=\xi-1$. Proposition 8.9(d) asserts that

$$H^S_{\xi-1}=\text{Coker}\left(\bar{\eta}_S:S\longrightarrow\mathfrak{R}(S)\right).$$

The general case follows from this special case by a chain of isomorphisms argument essentially identical to that used for the third value in part (v). □

CHAPTER 9

Examples of H_*^S

In this chapter, the results of the previous chapter are applied to the special cases $S = \langle \mathbb{Z} \rangle$, R, and L needed in our computations. By comparing H_*^R and H_*^L, we establish a connection between these two H_*-modules which plays an important role in the computations carried out in Chapter 5. We begin with $\langle \mathbb{Z} \rangle$ since this is especially simple.

PROPOSITION 9.1. *Let B be an abelian group. Then*

$$H_\omega^{\langle B \rangle} = \begin{cases} \langle B \rangle & \text{if } |\omega^G| = 0, \\ 0 & \text{otherwise.} \end{cases}$$

PROOF. The vanishing of $H_\omega^{\langle B \rangle}$ for $|\omega^G| \neq 0$ follows from the fact that, for any Mackey functor S, every value of H_*^S off of the vertical axis is constructed from $S(G/e)$ via some additive functor. The values on the vertical axis follow from Table 1.1 and part (i) of Theorem 8.1. □

The values of H_*^R and H_*^L depend critically on whether $p = 2$. The case $p \neq 2$ is simpler, and so is treated first.

PROPOSITION 9.2. *Assume that $p \neq 2$. Then*

$$H_\omega^R = \begin{cases} R & \text{if } |\omega| = 0 \text{ and } |\omega^G| \geq 0, \\ L & \text{if } |\omega| = 0 \text{ and } |\omega^G| < 0, \\ \langle \mathbb{Z}/p \rangle & \text{if } \begin{cases} |\omega| < 0,\ |\omega^G| \geq 0, \text{ and } |\omega^G| \text{ is even} \\ \text{or} \\ |\omega| > 0,\ |\omega^G| \leq -3, \text{ and } |\omega^G| \text{ is odd,} \end{cases} \\ 0 & \text{otherwise.} \end{cases}$$

and

$$H_\omega^L = \begin{cases} R & \text{if } |\omega| = 0 \text{ and } |\omega^G| > 0, \\ L & \text{if } |\omega| = 0 \text{ and } |\omega^G| \leq 0, \\ \langle \mathbb{Z}/p \rangle & \text{if } \begin{cases} |\omega| < 0,\ |\omega^G| > 0, \text{ and } |\omega^G| \text{ is even} \\ \text{or} \\ |\omega| > 0,\ |\omega^G| \leq -1, \text{ and } |\omega^G| \text{ is odd,} \end{cases} \\ 0 & \text{otherwise.} \end{cases}$$

PROOF. The values of H_*^R and H_*^L at the origin follow from Table 1.1. Both R and L satisfy the hypotheses of parts (a), (c) and (e) of Lemma 8.3. Moreover, L satisfies the hypotheses of part (b) of that lemma, and R satisfies the hypotheses of part (d). These observations give a number of vanishing results for H_*^R and H_*^L. Most of the still undetermined values of H_*^R and H_*^L follow from the observation that, if S is either R or L, then the map $\bar{\eta}_S \circ \bar{\epsilon}_S : \mathfrak{L}(S) \longrightarrow \mathfrak{R}(S)$ is just the canonical map $\Upsilon : L \longrightarrow R$. This map is a monomorphism with cokernel $\langle \mathbb{Z}/p \rangle$. The map $\bar{\eta}_L : L \longrightarrow \mathfrak{R}(L)$ can also be identified with Υ. This gives the one remaining value of H_*^L. Since $R = \mathfrak{R}(A)$, Remark 8.2(c) implies that the map $\widehat{\tau}_R : R_{G/e} \longrightarrow R$ has the same cokernel as Υ. The remaining uncomputed value of H_*^R follows from this. $\square$

The values of H_*^R and H_*^L are best visualized by plotting them in the plane, as in Figures 9.1 and 9.2 below.

Observe from these figures that the plot of H_*^L can be obtained simply by shifting the plot of H_*^R two units to the right. One way to say this is that, for any nontrivial irreducible complex G-representation ξ, H_*^L and $\Sigma^{2-\xi} H_*^R$ are isomorphic, at least as $RO(G)$-graded Mackey functors. In fact, a much stronger result holds.

					$\lvert\omega\rvert$			
					$\wedge$			
					$\vdots$			
	$\vdots$		$\vdots$		0			
$\cdots$	$\langle \mathbb{Z}/p \rangle$		$\langle \mathbb{Z}/p \rangle$					
					0			
$\cdots$	$\langle \mathbb{Z}/p \rangle$		$\langle \mathbb{Z}/p \rangle$					
$\cdots$		L		L	R	R	R	$\cdots > \lvert\omega^G\rvert$
					$\langle \mathbb{Z}/p \rangle$	$\langle \mathbb{Z}/p \rangle$	$\langle \mathbb{Z}/p \rangle$	$\cdots$
					$\langle \mathbb{Z}/p \rangle$	$\langle \mathbb{Z}/p \rangle$	$\langle \mathbb{Z}/p \rangle$	$\cdots$
					$\vdots$	$\vdots$	$\vdots$	

FIGURE 9.1. H_*^R for p odd

COROLLARY 9.3. *Assume that $p \neq 2$. Let ξ be any nontrivial irreducible complex G-representation, and let HA, HL, and HR be the equivariant Eilenberg-Mac Lane spectra representing equivariant ordinary homology with A, L, and R coefficients, respectively. Then HL and $\Sigma^{2-\xi} HR$ are equivalent in the equivariant stable category as module spectra over HA. Thus, H_*^L and $\Sigma^{2-\xi} H_*^R$ are isomorphic H_*-modules.*

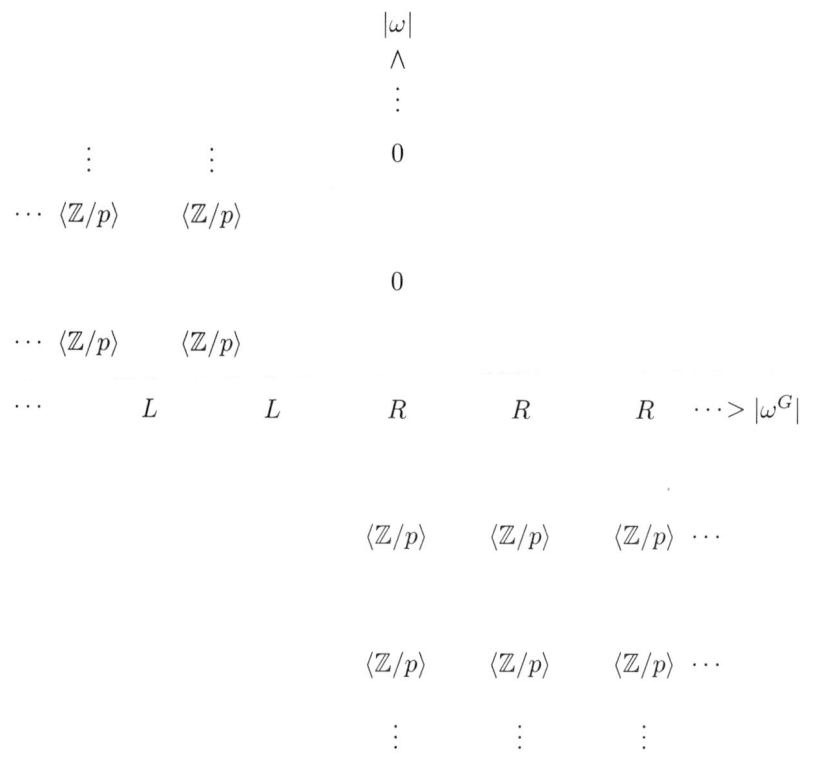

PROOF. Proposition 9.2 asserts that the equivariant stable homotopy "groups" $\pi^G_n(HL)$ and $\pi^G_n(\Sigma^{2-\xi}HR)$ are isomorphic Mackey functors for any integer n. Thus, $\Sigma^{2-\xi}HR$ is an equivariant Eilenberg-Mac Lane spectrum with $\pi^G_0(\Sigma^{2-\xi}HR) = L$. The uniqueness of equivariant Eilenberg-Mac Lane spectra, as module spectra over HA, follows easily from Proposition 5.4 of [**15**]. The claim about H^L_* and $\Sigma^{2-\xi}H^R_*$ is an obvious consequence of this. $\square$

						$\lvert\omega\rvert$			
						$\wedge$			
						$\vdots$			
	$\vdots$		$\vdots$			0			
$\cdots$	$\langle\mathbb{Z}/p\rangle$		$\langle\mathbb{Z}/p\rangle$		$\langle\mathbb{Z}/p\rangle$				
						0			
$\cdots$	$\langle\mathbb{Z}/p\rangle$		$\langle\mathbb{Z}/p\rangle$		$\langle\mathbb{Z}/p\rangle$				
$\cdots$		L		L		L	R	R	$\cdots > \lvert\omega^G\rvert$
						0	$\langle\mathbb{Z}/p\rangle$	$\langle\mathbb{Z}/p\rangle$	$\cdots$
						0	$\langle\mathbb{Z}/p\rangle$	$\langle\mathbb{Z}/p\rangle$	$\cdots$
						$\vdots$	$\vdots$	$\vdots$	

FIGURE 9.2. H^L_* for p odd

To compute H^S_* for $p = 2$, we must first compute the Mackey functors

$$\mathcal{C}(S) = \operatorname{Coker}(\widehat{\rho}_S : S \longrightarrow S_{G/e}) \qquad \text{and} \qquad \mathcal{K}(S) = \operatorname{Ker}(\widehat{\tau}_S : S_{G/e} \longrightarrow S)$$

introduced at the beginning of Chapter 8. The following result identifies these Mackey functors in some important special cases.

LEMMA 9.4. *Let $p = 2$.*

(a) *If B is an abelian group with a $\mathbb{Z}/2$-action, then there are natural isomorphisms*

$$\mathcal{C}(L(B)) \cong L_-(B) \qquad \textit{and} \qquad \mathcal{K}(R(B)) \cong R_-(B).$$

(b) *The canonical map $\bar{\eta}_A : A \longrightarrow R$ induces an isomorphism*

$$\mathcal{C}(A) \longrightarrow \mathcal{C}(R).$$

Moreover, $\mathcal{C}(R) \cong R_-$.

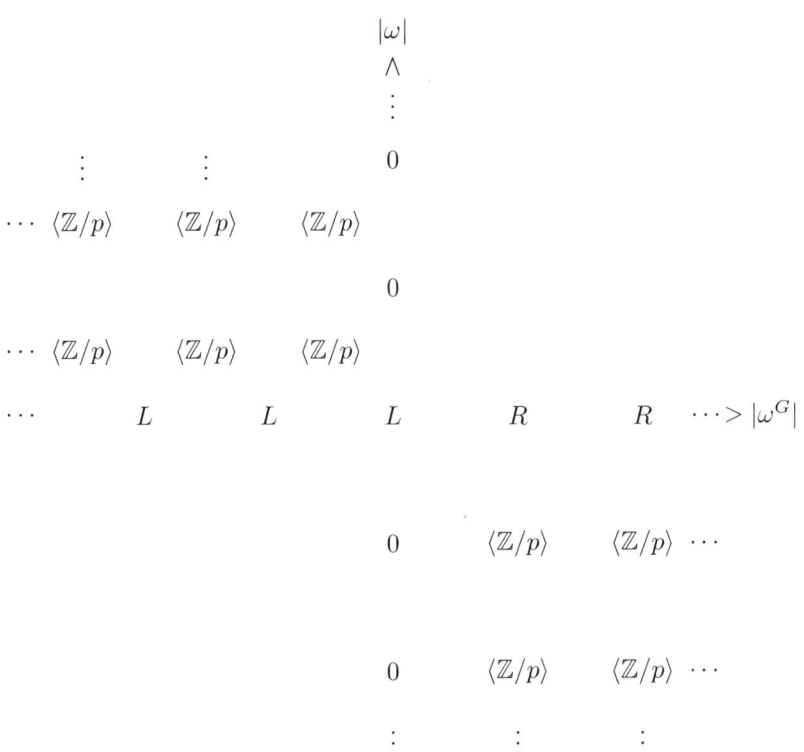

(c) *The canonical maps* $\bar{\epsilon}_A : L \longrightarrow A$ *and* $\bar{\eta}_A \circ \bar{\epsilon}_A : L \longrightarrow R$ *induce isomorphisms*

$$\mathcal{K}(L) \longrightarrow \mathcal{K}(A) \qquad \textit{and} \qquad \mathcal{K}(L) \longrightarrow \mathcal{K}(R).$$

Thus, $\mathcal{K}(A)$*,* $\mathcal{K}(L)$*, and* $\mathcal{K}(R)$ *are all isomorphic to* R_-*.*

PROOF. We want to think of B as having two $\mathbb{Z}/2$-actions — its original action and an alternative action coming from the identification of B with $B \otimes \mathbb{Z}_-$, on which $\mathbb{Z}/2$ acts diagonally via the original action on B and the sign action on $\mathbb{Z}_-$. To reduce confusion, we denote B with this alternative action by B_-. Observe that $L_-(B)$ and $R_-(B)$ are just $L(B_-)$ and $R(B_-)$. Denote the generaor of $\mathbb{Z}/2$ by σ, and let $\mathbb{Z}/2$ act on $B \oplus B$ by $\sigma(b, b') = (\sigma b', \sigma b)$. The usual diagonal and folding maps $\Delta : B \longrightarrow B \oplus B$ and $\nabla : B \oplus B \longrightarrow B$ are $\mathbb{Z}/2$-maps. Moreover, the "signed" diagonal and folding maps given by

$$\Delta_-(b) = (b, -b) \qquad \text{and} \qquad \nabla_-(b, b') = b - b'.$$

are $\mathbb{Z}/2$-maps when regarded as maps $\Delta_- : B_- \longrightarrow B \oplus B$ and $\nabla_- : B \oplus B \longrightarrow B_-$.

It is easy to check that the sequences

$$0 \longrightarrow B \xrightarrow{\Delta} B \oplus B \xrightarrow{\nabla_-} B_- \longrightarrow 0$$

and

$$0 \longrightarrow B_- \xrightarrow{\Delta_-} B \oplus B \xrightarrow{\nabla} B \longrightarrow 0$$

are short exact. The adjunctions defining L and R provide canonical maps

$$L(B \oplus B) \longrightarrow L(B)_{G/e} \qquad \text{and} \qquad R(B)_{G/e} \longrightarrow R(B \oplus B).$$

These two maps are isomorphisms, essentially because each is an isomorphism at G/e and all of the Mackey functors involved satisfy an appropriate form of induction with respect to the trivial subgroup. Under these isomorphisms, the maps $L(\Delta)$ and $R(\nabla)$ are identified with $\widehat{\rho}_{L(B)} : L(B) \longrightarrow L(B)_{G/e}$ and $\widehat{\tau}_{R(B)} : R(B)_{G/e} \longrightarrow R(B)$, respectively. Part (a) now follows from the two short exact sequences because L, being a left adjoint, preserves cokernels, and R, being a right adjoint, preserves kernels.

When evaluated at G/e, the exact sequence

$$R \xrightarrow{\widehat{\rho}_R} R_{G/e} \longrightarrow \mathcal{C}(R) \longrightarrow 0$$

defining $\mathcal{C}(R)$ can be identified with the special case of the first of the two short exact sequences above in which $B = \mathbb{Z}$. From this and the fact that the map $\widehat{\rho}_R$ is an isomorphism at G/G, it follows easily that $\mathcal{C}(R) \cong R_-$. For the rest of part (b), consider the commuting diagram

$$\begin{array}{ccccc} A & \xrightarrow{\widehat{\rho}_A} & A_{G/e} & \longrightarrow & \mathcal{C}(A) \\ {\scriptstyle \bar{\eta}_A}\downarrow & & \downarrow{\scriptstyle (\bar{\eta}_A)_{G/e}} & & \downarrow \\ R & \xrightarrow{\widehat{\rho}_R} & R_{G/e} & \longrightarrow & \mathcal{C}(R) \end{array}$$

in which the right vertical map is the one asserted to be an isomorphism. Its existence is ensured by the commutativity of the left square. Note that the left vertical arrow is an epimorphism and the middle vertical arrow is an isomorphism. From this, it follows that the right vertical arrow is an isomorphism.

For the first isomorphism in part (c), consider the commuting diagram

$$\begin{array}{ccccc} \mathcal{K}(L) & \longrightarrow & L_{G/e} & \xrightarrow{\widehat{\tau}_L} & L \\ \downarrow & & \downarrow{\scriptstyle (\bar{\epsilon}_A)_{G/e}} & & \downarrow{\scriptstyle \bar{\epsilon}_A} \\ \mathcal{K}(A) & \longrightarrow & A_{G/e} & \xrightarrow{\widehat{\tau}_A} & A \end{array}$$

in which the left vertical map is the one asserted to be an isomorphism. Its existence is ensured by the commutativity of the right square. Again, the middle vertical arrow is an isomorphism. The left vertical arrow is therefore an isomorphism since the right vertical arrow is a monomorphism. The second isomorphism in part (c) follows from the analogous diagram in which A is replaced by R and the map $\bar{\epsilon}_A$ is replaced by the composite $\bar{\eta}_A \circ \bar{\epsilon}_A$. The final assertion in part (c) follows from part (a) since $R = R(\mathbb{Z})$ if $\mathbb{Z}$ is given trivial $\mathbb{Z}/2$-action. □

The arguments used to prove Proposition 9.2 suffice to determine all the values of H_*^R and H_*^L for $p = 2$ except those for which $|\omega| = 0$ and $|\omega^G| = \pm 1$. Those missing values come from Theorem 8.1 and Lemma 9.4.

PROPOSITION 9.5. *Assume that $p = 2$. Then*

$$H_\omega^R = \begin{cases} R & \text{if } |\omega| = 0, |\omega^G| \geq 0, \text{ and } |\omega^G| \text{ is even,} \\ R_- & \text{if } |\omega| = 0, |\omega^G| \geq -1, \text{ and } |\omega^G| \text{ is odd,} \\ L & \text{if } |\omega| = 0, |\omega^G| < 0, \text{ and } |\omega^G| \text{ is even,} \\ L_- & \text{if } |\omega| = 0, |\omega^G| \leq -3, \text{ and } |\omega^G| \text{ is odd,} \\ \langle \mathbb{Z}/2 \rangle & \text{if } \begin{cases} |\omega| < 0, \ |\omega^G| \geq 0, \text{ and } |\omega^G| \text{ is even} \\ \quad \text{or} \\ |\omega| > 0, \ |\omega^G| \leq -3, \text{ and } |\omega^G| \text{ is odd,} \end{cases} \\ 0 & \text{otherwise.} \end{cases}$$

and

$$H_\omega^L = \begin{cases} R & \text{if } |\omega| = 0, |\omega^G| > 0, \text{ and } |\omega^G| \text{ is even,} \\ R_- & \text{if } |\omega| = 0, |\omega^G| > 0, \text{ and } |\omega^G| \text{ is odd,} \\ L & \text{if } |\omega| = 0, |\omega^G| \leq 0, \text{ and } |\omega^G| \text{ is even,} \\ L_- & \text{if } |\omega| = 0, |\omega^G| < 0, \text{ and } |\omega^G| \text{ is odd,} \\ \langle \mathbb{Z}/2 \rangle & \text{if } \begin{cases} |\omega| < 0, \ |\omega^G| > 0, \text{ and } |\omega^G| \text{ is even} \\ \quad \text{or} \\ |\omega| > 0, \ |\omega^G| \leq -1, \text{ and } |\omega^G| \text{ is odd,} \end{cases} \\ 0 & \text{otherwise.} \end{cases}$$

As in the case where $p \neq 2$, The values of H_*^R and H_*^L for $p = 2$ are best visualized by plotting them in the plane. These plots are given in Figures 9.3 and 9.4 below. Note that, as in the $p \neq 2$ case, the plot for H_*^L can be obtained by shifting the plot of H_*^R two units to the right. This motivates the following result:

COROLLARY 9.6. *Assume that $p = 2$. Let ζ be the one-dimensional real sign representation of $\mathbb{Z}$, and let HA, HL, $H(L_-)$, HR, and $H(R_-)$ be the equivariant Eilenberg-Mac Lane spectra representing equivariant ordinary homology with A, L,*

L_-, R, and R_- coefficients, respectively. Then there are equivalences

$$H(R_-) \simeq \Sigma^{1-\zeta} HR \qquad HL \simeq \Sigma^{2-2\zeta} HR \qquad H(L_-) \simeq \Sigma^{3-3\zeta} HR$$

of module spectra over HA in the equivariant stable category. These equivalences yield isomorphisms

$$H_*^{R-} \cong \Sigma^{1-\zeta} H_*^R \qquad H_*^L \cong \Sigma^{2-2\zeta} H_*^R \qquad H_*^{L-} \cong \Sigma^{3-3\zeta} H_*^R$$

of H_*-modules.

PROOF. As in the case $p \neq 2$, these results are obtained by computing the stable homotopy Mackey functors $\pi_n^G(\Sigma^{m-m\zeta} HR)$ for $1 \leq m \leq 3$ and $n \in \mathbb{Z}$. From this, it follows that $\Sigma^{m-m\zeta} HR$ is an equivariant Eilenberg-Mac Lane spectrum of the indicated type if $1 \leq m \leq 3$. □

$$
\begin{array}{cccccccccccc}
 & & & & & & |\omega| & & & & & \\
 & & & & & & \wedge & & & & & \\
 & \vdots & & \vdots & & & \vdots & & & & & \\
\cdots & \langle\mathbb{Z}/2\rangle & & \langle\mathbb{Z}/2\rangle & & & 0 & & & & & \\
\cdots & \langle\mathbb{Z}/2\rangle & & \langle\mathbb{Z}/2\rangle & & & 0 & & & & & \\
\cdots & L_- & L & L_- & L & R_- & R & R_- & R & R_- & R & \cdots > |\omega^G| \\
 & & & & & & \langle\mathbb{Z}/2\rangle & & \langle\mathbb{Z}/2\rangle & & \langle\mathbb{Z}/2\rangle & \cdots \\
 & & & & & & \langle\mathbb{Z}/2\rangle & & \langle\mathbb{Z}/2\rangle & & \langle\mathbb{Z}/2\rangle & \cdots \\
 & & & & & & \vdots & & \vdots & & \vdots &
\end{array}
$$

FIGURE 9.3. H_*^R for $p = 2$

$$
\begin{array}{cccccccccccc}
 & & & & & & |\omega| & & & & & \\
 & & & & & & \wedge & & & & & \\
 & \vdots & & \vdots & & & \vdots & & & & & \\
\cdots & \langle\mathbb{Z}/2\rangle & & \langle\mathbb{Z}/2\rangle & & \langle\mathbb{Z}/2\rangle & 0 & & & & & \\
\cdots & \langle\mathbb{Z}/2\rangle & & \langle\mathbb{Z}/2\rangle & & \langle\mathbb{Z}/2\rangle & 0 & & & & & \\
\cdots & L_- & L & L_- & L & L_- & L & R_- & R & R_- & R & \cdots > |\omega^G| \\
 & & & & & & 0 & & \langle\mathbb{Z}/2\rangle & & \langle\mathbb{Z}/2\rangle & \cdots \\
 & & & & & & 0 & & \langle\mathbb{Z}/2\rangle & & \langle\mathbb{Z}/2\rangle & \cdots \\
 & & & & & & \vdots & & \vdots & & \vdots &
\end{array}
$$

FIGURE 9.4. H_*^L for $p = 2$

CHAPTER 10

$RO(G)$-graded box products

Here, we introduce the box product on the category of modules over H_*. This construction lies at the heart of the weak form of the Universal Coefficient Theorem invoked in Chapter 5. We begin by introducing the category $\mathcal{M}_*$ of $RO(G)$-graded Mackey functors. This category is an obvious generalization of the category of $\mathbb{Z}$-graded abelian groups. The basic properties of $\mathcal{M}_*$ are therefore presented rather tersely, with fuller commentary only at the points where its behavior is not what one might expect from the nonequivariant case. The homology H_* of a point is a ring object in $\mathcal{M}_*$. Thus, from $\mathcal{M}_*$, we can construct the category H_*-Mod of modules over H_*. The bulk of this chapter is devoted to establishing the properties of this category which are needed for the proof of our weak Universal Coefficient Theorem. Since restricting to the case $G = \mathbb{Z}/p$ would save very little effort here, G is assumed to be an arbitrary finite group throughout this chapter.

Recall from [**11**], or Section V.9 of [**18**], that a Mackey functor C may be regarded as an additive functor from a small additive category $\mathcal{B}_G$, called the Burnside category, to the category Ab of abelian groups. The objects of the category $\mathcal{B}_G$ are finite G-sets. Also recall that, if C is a Mackey functor and X is a finite G-set, then the Mackey functor C_X is given on a G-set Y by

$$C_X(Y) = C(X \times Y).$$

This is a generalization of the $M_{G/e}$ construction introduced in Section 1.1. The category $\mathcal{M}$ of Mackey functors is a bicomplete symmetric monoidal closed abelian category which has enough projectives and injectives and satisfies $AB5$.

Definition 10.1. (a) An $RO(G)$-graded Mackey functor M is a collection $\{M_\alpha\}$ of Mackey functors indexed on the set $RO(G)$. A map $f : M \longrightarrow N$ of $RO(G)$-graded Mackey functors is the obvious collection of maps of Mackey functors. The category of $RO(G)$-graded Mackey functors for the group G is denoted $\mathcal{M}_*$.

(b) Let α be an element of $RO(G)$. The functor $\mathfrak{e}_\alpha$ from the category $\mathcal{M}_*$ to the category $\mathcal{M}$ of Mackey functors sends the $RO(G)$-graded Mackey functor M to its value M_α at α. The functor $\mathfrak{c}_\alpha$ from $\mathcal{M}$ to $\mathcal{M}_*$ sends a Mackey functor C to the $RO(G)$-graded Mackey functor whose values are C at α and zero at all other elements of $RO(G)$.

(c) For α an element of $RO(G)$, the functor $\Sigma^\alpha : \mathcal{M}_* \longrightarrow \mathcal{M}_*$ sends an $RO(G)$-graded Mackey functor M to the $RO(G)$-graded Mackey functor $\Sigma^\alpha M$ given by

$$(\Sigma^\alpha M)_\beta = M_{\beta-\alpha},$$

for $\beta \in RO(G)$.

(d) If M is an $RO(G)$-graded Mackey functor and X is a finite G-set, then the $RO(G)$-graded Mackey functor M_X is given at $\alpha \in RO(G)$ by

$$(M_X)_\alpha = (M_\alpha)_X \, .$$

(e) If M and N are $RO(G)$-graded Mackey functors, then the $RO(G)$-graded Mackey functor $M \square_* N$ is given at $\gamma \in RO(G)$ by

$$(M \square_* N)_\gamma = \bigoplus_{\alpha+\beta=\gamma} M_\alpha \square N_\beta,$$

where the box product on the right is in the category $\mathcal{M}$ of Mackey functors (see [**11, 15**]). Also, the $RO(G)$-graded Mackey functor $\langle M, N\rangle_*$ is given at $\gamma \in RO(G)$ by

$$(\langle M, N\rangle_*)_\gamma = \prod_\alpha \langle M_\alpha, N_{\alpha+\gamma}\rangle,$$

where the construction $\langle ?, ?\rangle$ on the right is the internal hom functor on the category $\mathcal{M}$ of Mackey functors (see [**11, 15**]).

The basic properties of the category $\mathcal{M}_*$ and the various constructions defined above are summarized in the next two propositions.

PROPOSITION 10.2. (a) *$\mathcal{M}_*$ is a bicomplete abelian category which has enough projectives and injectives and satisfies $AB5$.*

(b) *$\mathcal{M}_*$ is enriched over the category $\mathcal{M}$. Moreover, it is tensored and cotensored over $\mathcal{M}$.*

(c) *The functors $?\square_*?$ and $\langle ?, ?\rangle_*$ provide $\mathcal{M}_*$ with a symmetric monoidal closed structure which is consistent with its enrichment over $\mathcal{M}$. The unit for the product operation $\square_*$ on $\mathcal{M}_*$ is $\mathfrak{c}_0(A)$, where A is the Burnside ring Mackey functor.*

PROOF. Most of part (a) follows trivially from the observation that, if $RO(G)$ is regarded as a discrete category, then $\mathcal{M}_*$ is just the category of functors from $RO(G)$ to $\mathcal{M}$. For any $\alpha \in RO(G)$, the functors $\mathfrak{e}_\alpha$ and $\mathfrak{c}_\alpha$ are related by two adjunctions described in Proposition 10.4(a) below. It follows easily from these adjunctions that, if C is a projective (or injective) Mackey functor, then $\mathfrak{c}_\alpha(C)$ is a projective (or injective) $RO(G)$-graded Mackey functor. These adjunctions also imply that objects of this form provide $\mathcal{M}_*$ with enough projectives and injectives.

If M and N are $RO(G)$-graded Mackey functors, then the Mackey functor-valued hom construction which enriches $\mathcal{M}_*$ over $\mathcal{M}$ is just

$$\mathfrak{e}_0(\langle M, N\rangle_*) = \prod_\alpha \langle M_\alpha, N_\alpha\rangle.$$

It is easy to see that, if C is a Mackey functor and M is an $RO(G)$-graded Mackey functor, then $\mathfrak{c}_0(C)\square_* M$ and $\langle \mathfrak{c}_0(C), M\rangle_*$ are the tensor and cotensor, respectively, of C with M. Of course,

$$(\mathfrak{c}_0(C) \square_* M)_\alpha = C \square M_\alpha \quad \text{and} \quad (\langle \mathfrak{c}_0(C), M\rangle_*)_\alpha = \langle C, M_\alpha\rangle.$$

The unit and associativity isomorphisms for $\mathcal{M}_*$ follow easily from those for $\mathcal{M}$. The adjunction between the constructions $?\square_*?$ and $\langle ?, ?\rangle_*$ is the obvious generalization of the corresponding adjunction for graded abelian groups. Thus, the only non-obvious part of the symmetric monoidal closed structure on $\mathcal{M}_*$ is the commutativity isomorphism, which — like the commutativity isomorphism used in homological algebra for the category of graded abelian groups — involves sign changes. Unfortunately, the sign changes needed here involve nontrivial units in the Burnside ring Mackey functor. These are discussed in Remark 10.3 below. The symmetric monoidal structure is consistent with the enrichment over $\mathcal{M}$ in the sense that the functors $?\,\Box_*\,?$ and $\langle ?, ?\rangle_*$, and their adjunction, are enriched over $\mathcal{M}$. This is actually a formal consequence of the definition of the enrichment, but is also easily checked directly. □

REMARK 10.3. The commutativity isomorphism for the product on $\mathcal{M}_*$ takes the summand $M_\alpha \Box N_\beta$ of $(M \Box_* N)_{\alpha+\beta}$ to the summand $N_\beta \Box M_\alpha$ of $(N \Box_* M)_{\alpha+\beta}$ by the commutativity isomorphism for the box product on $\mathcal{M}$ composed with a "sign" change. Naively, one might expect the "sign" change to be given by multiplication by $(-1)^{|\alpha||\beta|}$, where $|\alpha|$ denotes the virtual dimension of α. If the order of G is odd, this naive approach to "sign" changes actually works because the Burnside ring of G contains no nontrivial units. However, for groups of even order (including $\mathbb{Z}/2$), the required "sign" is multiplication by a unit of the Burnside ring which need not be ± 1. The appropriate signs are given by a symmetric bilinear map

$$\mathrm{sgn} : RO(G) \times RO(G) \longrightarrow A(G)^\times,$$

where $A(G)^\times$ is the group of units of the Burnside ring of G. To define sgn, it suffices to specify $\mathrm{sgn}(V, W)$ when V and W are irreducible G-representations. If V and W are non-isomorphic irreducible representations, then $\mathrm{sgn}(V, W) = 1$. The element $\mathrm{sgn}(V, V)$ of $A(G)^\times$ is best described by thinking of it as an equivariant stable map $S^0 \longrightarrow S^0$. It is the map obtained by stabilizing the multiplication by -1 map from S^V to itself.

PROPOSITION 10.4. *Let C and D be Mackey functors, M and N be $RO(G)$-graded Mackey functors, α and β be elements of $RO(G)$, and X and Y be finite G-sets. Then*

(a) *The functors $\mathfrak{e}_\alpha : \mathcal{M}_* \longrightarrow \mathcal{M}$ and $\mathfrak{c}_\alpha : \mathcal{M} \longrightarrow \mathcal{M}_*$ are both enriched over $\mathcal{M}$. Moreover, $\mathfrak{e}_\alpha$ is both left adjoint and right adjoint to $\mathfrak{c}_\alpha$, and these two adjunctions are enriched over $\mathcal{M}$.*

(b) *There is a natural isomorphism*

$$\mathfrak{c}_\alpha(C) \Box_* \mathfrak{c}_\beta(D) \cong \mathfrak{c}_{\alpha+\beta}(C \Box D).$$

Thus, $\mathfrak{c}_0$ is a strict monoidal functor and $\mathfrak{e}_0$ is both a lax monoidal and a lax comonoidal functor.

(c) *The functor $\Sigma^\alpha : \mathcal{M}_* \longrightarrow \mathcal{M}_*$ is enriched over $\mathcal{M}$, and is an enriched adjoint equvalence with inverse $\Sigma^{-\alpha}$.*

(d) *There are natural isomorphisms*

$$(\Sigma^\alpha M) \Box_* (\Sigma^\beta N) \cong \Sigma^{\alpha+\beta}(M \Box_* N)$$

and

$$\langle \Sigma^\alpha M, \Sigma^\beta N\rangle_* \cong \Sigma^{\beta-\alpha}\langle M, N\rangle_*.$$

(e) *The endofunctor of $\mathcal{M}_*$ sending M to M_X is a self-adjoint functor enriched over $\mathcal{M}$.*

(f) *There are natural isomorphisms*

$$(M_X) \Box_* (N_Y) \cong (M \Box_* N)_{X\times Y}$$

and

$$\langle M_X, N_Y\rangle_* \cong (\langle M, N\rangle_*)_{X\times Y}.$$

For our purposes, the basic connection between the category $\mathcal{M}_*$ and equivariant homology is given by the following result.

PROPOSITION 10.5. *Let T be a commutative ring spectrum, and Y be a module spectrum over T in the equivariant stable homotopy category. Then the homology T_* of a point with respect T is a commutative ring object in $\mathcal{M}_*$, and the homology Y_* of a point with respect Y is a module over T_* in $\mathcal{M}_*$.*

In order to prove our weak Universal Coefficient Theorem, we need several results about the category of modules over H_*. However, everything we need to know about this category is true in the broader context of the category of modules over any commutative ring object in $\mathcal{M}_*$. Thus, for the rest of this chapter, T_* is a commutative ring object in the category $\mathcal{M}_*$. A module over T_* is an $RO(G)$-graded Mackey functor M together with a map $\zeta : T_* \square_* M \longrightarrow M$ for which the obvious diagrams commute. The category of T_* modules is denoted T_*-Mod. This category inherits all the good properties of the category $\mathcal{M}_*$.

DEFINITION 10.6. Let M and N be modules over T_*, and let K be an $RO(G)$-graded Mackey functor.

(a) The box product $M \square_{T_*} N$ is given by the coequalizer diagram

$$M \square_* T_* \square_* N \rightrightarrows M \square_* N \longrightarrow M \square_{T_*} N$$

in which the parallel arrows come from the actions of T_* on M and N.

(b) The internal hom functor $\langle M, N\rangle_{T_*}$ is given by the equalizer diagram

$$\langle M, N\rangle_{T_*} \longrightarrow \langle M, N\rangle_* \rightrightarrows \langle T_* \square_* M, N\rangle_* \cong \langle M, \langle T_*, N\rangle_*\rangle_*$$

in which the parallel arrows come from the actions of T_* on M and N.

(c) The $RO(G)$-graded Mackey functors $T_* \square_* K$ and $\langle T_*, K\rangle_*$ carry T_*-module structures derived from the action of T_* on itself. These two constructions are called the free and cofree T_*-modules associated to K.

PROPOSITION 10.7. (a) *The category T_*-Mod is a bicomplete abelian category having enough projectives and injectives and satisfying $AB5$.*

(b) *The category T_*-Mod is enriched over the category $\mathcal{M}$. Moreover, it is tensored and cotensored over $\mathcal{M}$.*

(c) *The functors $? \square_{T_*} ?$ and $\langle ?, ?\rangle_{T_*}$ provide T_*-Mod with a symmetric monoidal closed structure which is consistent with its enrichment over $\mathcal{M}$. The unit for the product operation $\square_{T_*}$ on T_*-Mod is T_*.*

(d) *The functors sending an $RO(G)$-graded Mackey functor K to $T_* \square_* K$ and $\langle T_*, K\rangle_*$ are left and right adjoint, respectively, to the forgetful functor from T_*-Mod to $\mathcal{M}_*$. All three of these functors are enriched over $\mathcal{M}$, and so are their associated adjunctions. The free T_*-module functor $T_* \square_* ?$ is strict monoidal, and the forgetful functor from T_*-Mod to $\mathcal{M}_*$ is lax monoidal.*

(e) *The constructions $\Sigma^\alpha M$, for $\alpha \in RO(G)$, and M_X, for a finite G-set X, restrict to endofunctors on T_*-Mod.*

(f) *Let* M *and* N *be* T_**-modules,* α *and* β *be elements of* $RO(G)$*, and* X *and* Y *be finite* G*-sets. Then there are natural isomorphisms*

$$(\Sigma^\alpha M) \,\square_{T_*}\, (\Sigma^\beta N) \cong \Sigma^{\alpha+\beta}(M \,\square_{T_*}\, N),$$
$$\langle \Sigma^\alpha M, \Sigma^\beta N\rangle_{T_*} \cong \Sigma^{\beta-\alpha}\langle M, N\rangle_{T_*},$$
$$(M_X) \,\square_{T_*}\, (N_Y) \cong (M \,\square_{T_*}\, N)_{X\times Y},$$

and

$$\langle M_X, N_Y\rangle_{T_*} \cong (\langle M, N\rangle_{T_*})_{X\times Y}$$

of T_**-modules.*

PROOF. The category T_*-Mod inherits limits and colimits from $\mathcal{M}_*$. Sufficient projectives for T_*-Mod are obtained by applying the free T_*-module construction to the projectives in $\mathcal{M}_*$. It follows from the appropriate adjunction in part (d) that these objects are, in fact, projective in T_*-Mod. Analogously, sufficient injectives for T_*-Mod are obtained by applying the cofree T_*-module construction to injectives in $\mathcal{M}_*$. If M and N are T_*-modules, then the Mackey functor-valued hom construction which enriches T_*-Mod over $\mathcal{M}$ is just $\mathfrak{e}_0(\langle M, N\rangle_{T_*})$. The tensors and cotensors for T_*-Mod are just the obvious restrictions to T_*-Mod of the analogous constructions on $\mathcal{M}_*$.

The unit, associativity, and commutativity isomorphisms for T_*-Mod are easily derived from the corresponding isomorphisms for $\mathcal{M}_*$. Similarly, the adjunction relating the functors $?\,\square_{T_*}?$ and $\langle ?, ?\rangle_{T_*}$ follows immediately from the analogous adjunction for $\mathcal{M}_*$.

The proof of part (d) is similar to a suitably formal proof of the corresponding result for ordinary commutative rings. Note that the lax monoidal structure on the forgetful functor from T_*-Mod to $\mathcal{M}_*$ is given by the unit map $u : \mathfrak{c}_0(A) \longrightarrow T_*$ and the canonical quotient map $M \,\square_*\, N \longrightarrow M \,\square_{T_*}\, N$. Part (e) is rather obvious, and part (f) follows directly from Propositions 10.4(d) and 10.4(f). □

CHAPTER 11

A weak Universal Coefficient Theorem

All that we need from the Universal Coefficient Theorem for the proof of our main results is the assertion that, if the $RO(G)$-graded ordinary homology $H^G_*(X;A)$ of a G-space X with Burnside ring coefficients is free over H_*, then, for certain Mackey functors M, the canonical map

$$\sigma^M_X : H^G_*(X;A) \square_{H_*} H^M_* \longrightarrow H^G_*(X;M)$$

is an isomorphism. Such a result, which we hereafter refer to as a weak Universal Coefficient Theorem, would clearly follow from any reasonable Universal Coefficient Theorem for $RO(G)$-graded ordinary homology. Unfortunately, since $RO(G)$-graded ordinary homology cannot be described in terms of chain complexes, obtaining such a theorem is not trivial. It is possible to extend the Universal Coefficient Theorems contained in [**4**] to G-spectra for any finite group G. Such an extension will appear in [**16**]. However, these results are applicable only to E_∞-ring G-spectra and their E_∞-module G-spectra. It is widely acknowledged that equivariant Eilenberg-MacLane spectra should carry such E_∞-structures, but a proof for this has not yet been published. To bridge this gap, we provide here an ad hoc proof of the weak Universal Coefficient Theorem in the cases for which we need it.

Recall the $\mathbb{Z}/p$-Mackey functors L, R and $\langle\mathbb{Z}\rangle$ introduced in Section 1.1.

PROPOSITION 11.1. *Let $G = \mathbb{Z}/p$, and let X be a G-space whose $RO(G)$-graded ordinary homology $H^G_*(X;A)$ with Burnside ring coefficients is free over H_*. Then the canonical maps*

$$\sigma^L_X : H^G_*(X;A) \square_{H_*} H^L_* \longrightarrow H^G_*(X;L),$$
$$\sigma^R_X : H^G_*(X;A) \square_{H_*} H^R_* \longrightarrow H^G_*(X;R),$$

and

$$\sigma^{\langle\mathbb{Z}\rangle}_X : H^G_*(X;A) \square_{H_*} H^{\langle\mathbb{Z}\rangle}_* \longrightarrow H^G_*(X;\langle\mathbb{Z}\rangle)$$

are isomorphisms.

Our proof of this result uses spectrum-level arguments. It is therefore convenient to break our usual convention and work with reduced, rather than unreduced, homology. Throughout the argument, the analogous results for unreduced homology can be obtained by replacing the G-space X by X_+.

To set the stage for the proof of this result, we begin with some observations applicable to any finite group G, any commutative ring G-spectrum T, and any module G-spectrum Y over T. Motivated by Definition 1.13, we say that the reduced T-homology $\widetilde{T}_*X$ of a G-space X is free over T_* if there is an isomorphism

$$\bigoplus_i \Sigma^{\omega_i}(T_*)_{G/H_i} \cong \widetilde{T}_*X$$

of T_*-modules for some collection $\{H_i\}$ of subgroups of G and some collection $\{\omega_i\}$ of elements of $RO(G)$. Given such an isomorphism, there are canonical maps

$$(G/H_i)_+ \wedge S^{\omega_i} \longrightarrow X \wedge T$$

which may be thought of as the free generators of $\widetilde{T}_*X$. Combining these generators, we obtain a map

$$f : \bigvee_i (G/H_i)_+ \wedge S^{\omega_i} \longrightarrow X \wedge T.$$

Let

$$\tilde{f}_Y : \bigvee_i (G/H_i)_+ \wedge S^{\omega_i} \wedge Y \longrightarrow X \wedge Y$$

be the composite

$$\bigvee_i (G/H_i)_+ \wedge S^{\omega_i} \wedge Y \xrightarrow{f \wedge 1} X \wedge T \wedge Y \xrightarrow{1 \wedge \zeta} X \wedge Y$$

in which ζ is the map giving the action of T on Y. It is easy to check that the map $\tilde{f}_Y$ is natural with respect to maps between T-modules. Denote the $RO(G)$-graded, Mackey-functor-valued stable homotopy "groups" of a G-spectrum Z by $\pi_*^G Z$. For any T-module Y, there is a canonical isomorphism

$$\bigoplus_i \Sigma^{\omega_i}(Y_*)_{G/H_i} \cong \pi_*^G\Big(\bigvee_i (G/H_i)_+ \wedge S^{\omega_i} \wedge Y\Big).$$

The original isomorphism identifying $\widetilde{T}_*X$ as a free T_*-module can be recovered as the composite

$$\bigoplus_i \Sigma^{\omega_i}(T_*)_{G/H_i} \cong \pi_*^G\Big(\bigvee_i (G/H_i)_+ \wedge S^{\omega_i} \wedge T\Big) \xrightarrow{(\tilde{f}_T)_*} \pi_*^G(X \wedge T) \cong \widetilde{T}_*X$$

of this canonical isomorphism and the map in homotopy induced by $\tilde{f}_T$. Thus, the map $\tilde{f}_T$ is a stable equivalence of G-spectra.

The connection between the maps $\tilde{f}_Y$ and the desired weak Universal Coefficient Theorem is described by the commuting diagram

$$\begin{array}{ccccc}
\big(\bigoplus_i \Sigma^{\omega_i}(T_*)_{G/H_i}\big) \,\square_{T_*}\, Y_* & & \xrightarrow{\cong} & & \bigoplus_i \Sigma^{\omega_i}(Y_*)_{G/H_i} \\
\| \wr & & & & \| \wr \\
\pi_*^G(\vee_i (G/H_i)_+ \wedge S^{\omega_i} \wedge T) \,\square_{T_*}\, Y_* & \searrow & & \nearrow & \pi_*^G(\vee_i (G/H_i)_+ \wedge S^{\omega_i} \wedge Y) \\
{\scriptstyle (\tilde{f}_T)_* \square_{T_*} 1}\downarrow \cong & & \pi_*^G(\vee_i (G/H_i)_+ \wedge S^{\omega_i} \wedge T \wedge Y) & & \downarrow {\scriptstyle (\tilde{f}_Y)_*} \\
\widetilde{T}_*X \,\square_{T_*}\, Y_* & & \xrightarrow[\sigma_X^Y]{} & & \widetilde{Y}_*X
\end{array}$$

in which the top isomorphism comes from Proposition 10.7(f). From the diagram, it follows that σ_X^Y is an isomorphism if and only if $\tilde{f}_Y$ is a stable equivalence of G-spectra.

Unfortunately, the most we can prove about $\tilde{f}_Y$ in general is the following:

LEMMA 11.2. *Let T be a commutative ring G-spectrum, Y be a module G-spectrum over T, and X be a G-space whose reduced T-homology is free. Then the map $\tilde{f}_Y$ is a split epimorphism in the equivariant stable category.*

PROOF. Let $\{H_i\}$ and $\{\omega_i\}$ be the collections of subgroups of G and elements of $RO(G)$, respectively, such that

$$\bigoplus_i \Sigma^{\omega_i}(T_*)_{G/H_i} \cong \widetilde{T}_* X.$$

Then the commuting diagram

$$\begin{array}{ccc}
\vee_i(G/H_i)_+ \wedge S^{\omega_i} \wedge T \wedge Y & \xrightarrow{1\wedge 1\wedge\zeta} & \vee_i(G/H_i)_+ \wedge S^{\omega_i} \wedge Y \\
{\scriptstyle \tilde{f}_T\wedge 1=\tilde{f}_{T\wedge Y}}\downarrow \cong & & \\
X \wedge T \wedge Y & \searrow^{1\wedge\zeta} & \downarrow \tilde{f}_Y \\
{\scriptstyle 1\wedge e\wedge 1}\uparrow & & \\
X \wedge S^0 \wedge Y & \xrightarrow[\cong]{} & X \wedge Y
\end{array}$$

displays the desired splitting. Here, $e : S^0 \longrightarrow T$ is the unit map for T. □

Since the composite

$$Y \cong S^0 \wedge Y \xrightarrow{e\wedge 1} T \wedge Y$$

is not a map of T-modules, it is not possible to give an analogous argument showing that $\tilde{f}_Y$ is a split monomorphism.

We now specialize to the case in which T is the Eilenberg-Mac Lane spectrum HA associated to the Burnside ring Mackey functor for some finite group G. In proving Proposition 11.1, we make use of a special property of modules over HA. Assume that

$$0 \longrightarrow M' \xrightarrow{j} M \xrightarrow{q} M'' \longrightarrow 0$$

is a short exact sequence of Mackey functors. Then the equivariant Eilenberg-Mac Lane spectra HM', HM, and HM'' are HA-module spectra and the induced maps $\bar{j} : HM' \longrightarrow HM$ and $\bar{q} : HM \longrightarrow HM''$ are HA-module maps in the equivariant stable category (see Proposition 5.4 of [**15**]). Moreover, the sequence

$$HM' \xrightarrow{\bar{j}} HM \xrightarrow{\bar{q}} HM''$$

is a cofibre sequence in the equivariant stable category. Let $\partial : HM'' \longrightarrow \Sigma HM'$ be the boundary map associated to this cofibre sequence.

LEMMA 11.3. *The boundary map $\partial : HM'' \longrightarrow \Sigma HM'$ is a map of HA-module spectra.*

PROOF. As noted in the proof of Lemma 2.2 of [**13**], the unit map $e : S^0 \longrightarrow HA$ of HA is the inclusion of its 0-skeleton, and no 1-cells are needed to form HA from S^0. Thus, the cofibre HA/S^0 of e is 1-connected. We wish to show that the diagram

$$\begin{array}{ccc}
HA \wedge HM'' & \xrightarrow{\zeta''} & HM'' \\
{\scriptstyle 1\wedge\partial}\downarrow & & \downarrow{\scriptstyle \partial} \\
HA \wedge \Sigma HM' & \xrightarrow{1\wedge\Sigma\zeta'} & \Sigma HM'
\end{array}$$

commutes in the equivariant stable category. This can be rephrased as the assertion that a certain element of the equivariant homotopy set $[HA \wedge HM'', \Sigma HM']_G$ vanishes. The image of this element under the map

$$[HA \wedge HM'', \Sigma HM']_G \xrightarrow{(e\wedge 1)^*} [S^0 \wedge HM'', \Sigma HM']_G$$

certainly vanishes. However, since $[(HA/S^0) \wedge HM'', \Sigma HM']_G$ is zero, $(e \wedge 1)^*$ is a monomorphism. □

In the nonequivariant context, this result suffices to give a weak Universal Coefficient Theorem for ordinary homology because the category of abelian groups has homological dimension one. However, for a nontrivial finite group G, most Mackey functors have infinite homological dimension.

Rather than attempt to deal with this homological difficulty, we now restrict our attention to the special case $G = \mathbb{Z}/p$. It is easy to see that the $\mathbb{Z}/p$-Mackey functor R is a Mackey functor ring. Thus, for any G-space X, $\widetilde{H}_*^G(X; R)$ consists of R-modules. The following lemma allows us to exploit this fact.

LEMMA 11.4. *Let M be a Mackey functor module over the Mackey functor ring R, and let C be an abelian group. Then any map $h : M \longrightarrow \langle C \rangle$ factors through $\langle C' \rangle$, where C' is the subgroup of C consisting of elements annihilated by p. Thus, if there is no p-torsion in C, then there are no nontrivial maps from M to $\langle C \rangle$.*

PROOF. Let ξ be the generator of $R(G/G)$. Recall that $p\xi = \tau(\rho(\xi))$, where τ and ρ are the transfer and restriction maps for R. Thus, for any $x \in M(G/G)$,

$$px = p\xi x = \tau(\rho(\xi))x = \tau(\rho(\xi x)).$$

The submodule $pM(G/G)$ of $M(G/G)$ is therefore contained in the image of the transfer $\tau : M(G/e) \longrightarrow M(G/G)$. It follows that $h(pM(G/G))$ must be zero, and so the image of $M(G/G)$ under h must be contained in C'. □

We can now prove our weak Universal Coefficient result.

PROOF OF PROPOSITION 11.1. We prove the analogous result for reduced homology. The asserted result for unreduced homology is then obtained by replacing X by X_+. Let X be a G-space whose reduced $RO(G)$-graded ordinary homology $\widetilde{H}_*^G(X; A)$ with Burnside ring coefficients is free over H_*. Also, let $\{H_i\}$ and $\{\omega_i\}$ be the collections of subgroups of G and elements of $RO(G)$, respectively, such that

$$\bigoplus_i \Sigma^{\omega_i}(H_*)_{G/H_i} \cong \widetilde{H}_*^G(X; A).$$

Recall that there is a short exact sequence

$$0 \longrightarrow \langle \mathbb{Z} \rangle \longrightarrow A \longrightarrow R \longrightarrow 0$$

of Mackey functors. From this short exact sequence, we obtain long exact sequences for the reduced homology of both X and $F = \vee_i (G/H_i)_+ \wedge S^{\omega_i}$. Lemma 11.3 implies that the maps $\tilde{f}_{H\langle\mathbb{Z}\rangle}$, $\tilde{f}_{HA}$, and $\tilde{f}_{HR}$ induce a map

$$\begin{array}{ccccccc}
\cdots \xrightarrow{\partial_F} & \widetilde{H}_*^G(F; \langle \mathbb{Z} \rangle) & \longrightarrow & \widetilde{H}_*^G(F; A) & \longrightarrow & \widetilde{H}_*^G(F; R) & \xrightarrow{\partial_F} \cdots \\
& \Big\downarrow {\scriptstyle (\tilde{f}_{H\langle\mathbb{Z}\rangle})_*} & & {\scriptstyle \cong}\Big\downarrow {\scriptstyle (\tilde{f}_{HA})_*} & & \Big\downarrow {\scriptstyle (\tilde{f}_{HR})_*} & \\
\cdots \xrightarrow{\partial_X} & \widetilde{H}_*^G(X; \langle \mathbb{Z} \rangle) & \longrightarrow & \widetilde{H}_*^G(X; A) & \longrightarrow & \widetilde{H}_*^G(X; R) & \xrightarrow{\partial_X} \cdots
\end{array}$$

between these two long exact sequences. By Lemma 11.2, the vertical maps in this diagram are split epimorphisms. Moreover, by assumption, $(\tilde{f}_{HA})_*$ is an isomorphism. For any $\omega \in RO(G)$, $\widetilde{H}^G_\omega(F; \langle\mathbb{Z}\rangle)$ is either zero or a sum of copies of $\langle\mathbb{Z}\rangle$. Thus, by Lemma 11.4, ∂_F must be zero. A simple diagram chase now gives that $(\tilde{f}_{H\langle\mathbb{Z}\rangle})_*$ is a monomorphism. Thus, $(\tilde{f}_{H\langle\mathbb{Z}\rangle})_*$ is a isomorphism, and the five lemma gives that $(\tilde{f}_{HR})_*$ is an isomorphism. Our observation about the relation between σ_X^Y and $\tilde{f}_Y$ for an arbitrary module spectrum Y then gives that $\sigma_X^{\langle\mathbb{Z}\rangle}$ and σ_X^R are isomorphisms. The fact that σ_X^L is also an isomorphism follows immediately from the connection between the equivariant Eilenberg-Mac Lane spectra associated to L and R described in Corollaries 9.3 and 9.6. □

CHAPTER 12

Observations about Mackey functors

This chapter supplies some facts about short exact sequences of $\mathbb{Z}/p$-Mackey functors which are used in Chapters 5 and 6. Most of these observations are the sort of thing that would be left to the reader if we were working in an abelian category more familiar than the category of Mackey functors.

LEMMA 12.1. *Let*

$$0 \longrightarrow A \xrightarrow{\ \iota\ } D \xrightarrow{\ \pi\ } \langle \mathbb{Z}/p \rangle \longrightarrow 0$$

be a short exact sequence of Mackey functors. Then

$$D \cong \begin{cases} A \oplus \langle \mathbb{Z}/p \rangle & \text{if the sequence splits,} \\ R \oplus \langle \mathbb{Z} \rangle & \text{otherwise.} \end{cases}$$

Moreover, if $D \cong R \oplus \langle \mathbb{Z} \rangle$*, then the two components* $\iota_1 : A \longrightarrow R$ *and* $\iota_2 : A \longrightarrow \langle \mathbb{Z} \rangle$ *of the map* ι *are surjective.*

PROOF. Clearly, $D(G/e) = \mathbb{Z}$, and $D(G/G)$ is either $\mathbb{Z} \oplus \mathbb{Z} \oplus \mathbb{Z}/p$ or $\mathbb{Z} \oplus \mathbb{Z}$. If $D(G/G) = \mathbb{Z} \oplus \mathbb{Z} \oplus \mathbb{Z}/p$, then the short exact sequence must split because the restriction map $\rho : D(G/G) \longrightarrow D(G/e)$ must vanish on the $\mathbb{Z}/p$ summand of $D(G/G)$. On the other hand, if $D(G/G) = \mathbb{Z} \oplus \mathbb{Z}$, then the short exact sequence obviously does not split. Thus, assume $D(G/G) = \mathbb{Z} \oplus \mathbb{Z}$. Let μ and $\bar{\tau}$ be the standard generators of $A(G/G)$. Also, let $z = \iota(\mu)$, $u = \rho(z)$, and $t = \iota(\bar{\tau}) = \tau(u)$. Note that u must generate $D(G/e) = \mathbb{Z}$. A simple rank argument indicates that the kernel of the restriction map $\rho : D(G/G) \longrightarrow D(G/e)$ is isomorphic to $\mathbb{Z}$. Let y be a generator of this kernel. Then $\pi(y) \neq 0$. Otherwise, select some $w \in D(G/G)$ such that $\pi(w) \neq 0$. There is an integer s such that $\rho(w - sz) = 0$, so that $w - sz = ry$ for some integer r. From this we get the contradiction that $\pi(w) = \pi(ry + sz) = 0$.

Since $\pi(y)$ isn't zero, t, y, and z generate $D(G/G)$. However, $\rho(t - pz) = 0$, so there is an integer b such that $t - pz = by$, from which it follows that y and z generate $D(G/G)$. Note that $\pi(by) = \pi(t - pz) = 0$, so p divides b. Thus, there is an integer b' such that $t = p(z + b'y)$. By replacing y by its negative if necessary, we can assume that $b' \geq 0$.

Observe that py is in the image of ι since $\pi(py) = 0$. In fact, because ι is injective, there is an integer c such that $py = c(t - pz)$. Substituting in for t, we have that $py = b'cpy$. Thus $b' = c = 1$ since $b' \geq 0$. Let $z' = y + z$. Then $t = pz'$, and $u = \rho(z')$. It follows that z' generators a copy of R contained in D. Further, y generates a copy of $\langle \mathbb{Z} \rangle$ in D, and it is easy to see that D is the direct sum of these two Mackey functors. Since $z = z' - y'$, the two maps ι_1 and ι_2 are surjective. □

LEMMA 12.2. (a) *Let*

$$0 \longrightarrow L \xrightarrow{\ \iota\ } D \xrightarrow{\ \pi\ } \langle \mathbb{Z} \rangle \longrightarrow 0 \tag{12.1}$$

be a short exact sequence of Mackey functors. Then

$$D \cong \begin{cases} L \oplus \langle \mathbb{Z} \rangle & \text{if the sequence splits,} \\ A[d] & \text{otherwise.} \end{cases}$$

Here, d is assumed to be relatively prime to p. Moreover, if $D \cong A[d]$, then d is determined in $\mathbb{Z}/p$ up to sign by the fact that there is an element $x \in D(G/G)$ such that $\pi(x)$ generates $\langle \mathbb{Z} \rangle(G/G) = \mathbb{Z}$ and $\rho(x) = d \in D(G/e) = \mathbb{Z}$.

(b) *Let*

$$0 \longrightarrow \langle \mathbb{Z} \rangle \xrightarrow{\iota} D \xrightarrow{\pi} R \longrightarrow 0 \tag{12.2}$$

be a short exact sequence of Mackey functors. Then

$$D \cong \begin{cases} \langle \mathbb{Z} \rangle \oplus R & \text{if the sequence splits,} \\ A[d] & \text{otherwise.} \end{cases}$$

Here, d is assumed to be relatively prime to p.

(c) *If the Mackey functor D fits into short exact sequences of both form* (12.1) *and form* (12.2), *then $D \cong A[d]$ for some integer d prime to p.*

PROOF. For part (a), note that $D(G/G)$ must be $\mathbb{Z} \oplus \mathbb{Z}$. Let $t' \in D(G/G)$ be $\iota(t)$, where t is the standard generator of $L(G/G)$. Then $\rho(t') = p \in D(G/e) = \mathbb{Z}$. Select an element x of $D(G/G)$ such that $\pi(x)$ generates $\langle \mathbb{Z} \rangle(G/G) = \mathbb{Z}$. Clearly t' and x generate $D(G/G)$. Let $d = \rho(x) \in D(G/e) = \mathbb{Z}$. If p divides d, then by adding some multiple of t' to x we can obtain an element x' of $D(G/G)$ such that $\pi(x)$ generates $\langle \mathbb{Z} \rangle(G/G)$ and $\rho(x') = 0$. In this case, the short exact sequence obviously splits. Thus, assume that $d \in D(G/e)$ is relatively prime to p. Any other element y of $D(G/G)$ such that $\pi(y)$ generates $\langle \mathbb{Z} \rangle(G/G)$ must be of the form $\pm x + at'$ for some integer a. Therefore, $\rho(y) \in D(G/e)$ is also relatively prime to p, and the short exact sequence cannot split. Moreover, it is easy to check that $D \cong A[d]$ by a map sending x and t' to the standard generators μ and τ of $A[d](G/G)$.

For part (b), again note that $D(G/G)$ must be $\mathbb{Z} \oplus \mathbb{Z}$. Let $k \in D(G/G)$ be the image of a generator of $\langle \mathbb{Z} \rangle(G/G) = \mathbb{Z}$. Then $\rho(k) = 0 \in D(G/e) = \mathbb{Z}$. Select an element y of $D(G/G)$ such that $\pi(y)$ generates $R(G/G) = \mathbb{Z}$. Clearly k and y generate $D(G/G)$. Moreover, $u = \rho(y)$ generates $D(G/e)$. It is easy to check that $\tau(u) = py - ak$ for some integer a. If p divides a, then we can adjust y by some multiple of k to obtain an element y' of $D(G/G)$ such that $\pi(y')$ generates $R(G/G)$, $u = \rho(y')$, and $\tau(u) = py'$. In this case, the short exact sequence obviously splits. Thus, assume that a is relatively prime to p. Let z be any other element of $D(G/G)$ such that $\pi(z)$ generates $R(G/G)$. Then $z = \pm y + bk$ for some integer b and $\rho(z) = \pm u$. Further, $\tau(u) = \pm pz - a'k$ for some integer a' which is congruent to a modulo p. But then a' is relatively prime to p, from which it follows that the exact sequence cannot split. Select an integer d such that ad is congruent to 1 modulo p. It is easy to check that $D \cong A[d]$ by a map sending y and k to the standard σ and κ generators of $A[d](G/G)$.

For part (c), it suffices to show that the Mackey functors $\langle \mathbb{Z} \rangle \oplus R$, $L \oplus \langle \mathbb{Z} \rangle$, and $A[d]$ are pairwise nonisomorphic. Clearly, $L \oplus \langle \mathbb{Z} \rangle$ is not isomorphic to either of the other two because the the restriction map is surjective in $\langle \mathbb{Z} \rangle \oplus R$ and $A[d]$, but not in $L \oplus \langle \mathbb{Z} \rangle$. Further, $\langle \mathbb{Z} \rangle \oplus R$ is not isomorphic to $A[d]$ because every element in $(\langle \mathbb{Z} \rangle \oplus R)(G/G)$ which is in the image of the transfer is p-divisible, whereas there are elements in the image of the tranfer in $A[d](G/G)$ which are not p-divisible. □

LEMMA 12.3. *For any nonzero map* $\pi : R \longrightarrow \langle \mathbb{Z}/p \rangle$, $\operatorname{Ker} \pi \cong L$.

PROOF. Let $K = \operatorname{Ker} \pi$. Clearly, $K(G/G) \cong \mathbb{Z} \cong K(G/e)$. The restriction and transfer maps for K are easily computed from the embedding of K into R. From this, it follows immediately that $K = L$. □

Bibliography

1. G. E. Bredon, *Equivariant cohomology theories*, Lecture Notes in Math., vol. 34, Springer, Berlin, 1967.
2. M. Cole, J. P. C. Greenlees, and I. Kriz, *The universality of equivariant complex bordism*, Math. Z. **239** (2002), no. 3, 455–475.
3. A. W. M. Dress, *Contributions to the theory of induced representations*, Algebraic K-theory, II: "Classical" algebraic K-theory and connections with arithmetic (Proc. Conf., Battelle Memorial Inst., Seattle, Wash., 1972), Lecture Notes in Math., vol. 342, Springer, Berlin, 1973, pp. 183–240.
4. A. D. Elmendorf, I. Kriz, M. A. Mandell, and J. P. May, *Rings, modules, and algebras in stable homotopy theory*, Mathematical Surveys and Monographs, vol. 47, American Mathematical Society, Providence, RI, 1997, (with an appendix by M. Cole).
5. H. Fausk, L. G. Lewis, Jr., and J. P. May, *The Picard group of equivariant stable homotopy theory*, Adv. Math. **163** (2001), no. 1, 17–33.
6. K. K. Ferland, *On the $RO(G)$-graded equivariant ordinary cohomology of generalized G-cell complexes for $G = \mathbb{Z}/p$*, Ph.D. thesis, Syracuse University, 1999.
7. J. A. Green, *Axiomatic representation theory for finite groups*, J. Pure Appl. Algebra **1** (1971), no. 1, 41–77.
8. P. Griffiths and J. Harris, *Principles of algebraic geometry*, John Wiley & Sons, 1978.
9. H. Hiller, *Geometry of Coxeter groups*, Research Notes in Mathematics, vol. 54, Pitman (Advanced Publishing Program), Boston, Mass., 1982.
10. S. Illman, *Equivariant singular homology and cohomology. I*, Mem. Amer. Math. Soc. **1**, issue 2 (1975), no. 156.
11. L. G. Lewis, Jr., *The theory of Green functors*, Mimeographed notes, 1980.
12. ______, *The $RO(G)$-graded equivariant ordinary cohomology of complex projective spaces with linear $\mathbb{Z}/p$ actions*, Algebraic topology and transformation groups, Proceedings, Göttingen 1987, Lecture Notes in Math., vol. 1361, Springer, Berlin, 1988, pp. 53–122.
13. ______, *The equivariant Hurewicz map*, Trans. Amer. Math. Soc. **329** (1992), no. 2, 433–472.
14. ______, *Change of universe functors in equivariant stable homotopy theory*, Fund. Math. **148** (1995), no. 2, 117–158.
15. ______, *The category of Mackey functors for a compact Lie group*, Group representations: cohomology, group actions and topology (Seattle, WA, 1996), Proc. Sympos. Pure Math., vol. 63, Amer. Math. Soc., Providence, RI, 1998, pp. 301–354.
16. L. G. Lewis, Jr. and M. A. Mandell, *Equivariant universal coefficient and Künneth spectral sequences*, In preparation.
17. L. G. Lewis, Jr., J. P. May, and J. E. McClure, *Ordinary $RO(G)$-graded cohomology*, Bull. Amer. Math. Soc. (N.S.) **4** (1981), no. 2, 208–212.
18. L. G. Lewis, Jr., J. P. May, and M. Steinberger, *Equivariant stable homotopy theory*, Lecture Notes in Math., vol. 1213, Springer, Berlin, 1986, (with contributions by J. E. McClure).
19. J. P. May, *Equivariant homotopy and cohomology theory*, CBMS Regional Conference Series in Mathematics, vol. 91, Published for the Conference Board of the Mathematical Sciences, Washington, DC, 1996.
20. J. W. Milnor and J. D. Stasheff, *Characteristic classes*, Annals of Mathematics Studies, vol. 76, Princeton University Press, 1974.
21. M. Y. Oruç, *The equivariant Steenrod algebra*, Topology Appl. **32** (1989), 77–108.
22. J. Thévenaz, *A visit to the kingdom of the Mackey functors*, Bayreuth. Math. Schr. **33** (1990), 215–241.
23. A. G. Wasserman, *Equivariant differential topology*, Topology **8** (1969), 127–150.

Editorial Information

To be published in the *Memoirs*, a paper must be correct, new, nontrivial, and significant. Further, it must be well written and of interest to a substantial number of mathematicians. Piecemeal results, such as an inconclusive step toward an unproved major theorem or a minor variation on a known result, are in general not acceptable for publication. Papers appearing in *Memoirs* are generally longer than those appearing in *Transactions*, which shares the same editorial committee.

As of October 1, 2003, the backlog for this journal was approximately 5 volumes. This estimate is the result of dividing the number of manuscripts for this journal in the Providence office that have not yet gone to the printer on the above date by the average number of monographs per volume over the previous twelve months, reduced by the number of volumes published in four months (the time necessary for preparing a volume for the printer). (There are 6 volumes per year, each containing at least 4 numbers.)

A Consent to Publish and Copyright Agreement is required before a paper will be published in the *Memoirs*. After a paper is accepted for publication, the Providence office will send a Consent to Publish and Copyright Agreement to all authors of the paper. By submitting a paper to the *Memoirs*, authors certify that the results have not been submitted to nor are they under consideration for publication by another journal, conference proceedings, or similar publication.

Information for Authors

Memoirs are printed from camera copy fully prepared by the author. This means that the finished book will look exactly like the copy submitted.

The paper must contain a *descriptive title* and an *abstract* that summarizes the article in language suitable for workers in the general field (algebra, analysis, etc.). The *descriptive title* should be short, but informative; useless or vague phrases such as "some remarks about" or "concerning" should be avoided. The *abstract* should be at least one complete sentence, and at most 300 words. Included with the footnotes to the paper should be the 2000 *Mathematics Subject Classification* representing the primary and secondary subjects of the article. The classifications are accessible from `www.ams.org/msc/`. The list of classifications is also available in print starting with the 1999 annual index of *Mathematical Reviews*. The Mathematics Subject Classification footnote may be followed by a list of *key words and phrases* describing the subject matter of the article and taken from it. Journal abbreviations used in bibliographies are listed in the latest *Mathematical Reviews* annual index. The series abbreviations are also accessible from `www.ams.org/publications/`. To help in preparing and verifying references, the AMS offers MR Lookup, a Reference Tool for Linking, at `www.ams.org/mrlookup/`. When the manuscript is submitted, authors should supply the editor with electronic addresses if available. These will be printed after the postal address at the end of the article.

Electronically prepared manuscripts. The AMS encourages electronically prepared manuscripts, with a strong preference for $\mathcal{A}_{\mathcal{M}}\mathcal{S}$-LaTeX. To this end, the Society has prepared $\mathcal{A}_{\mathcal{M}}\mathcal{S}$-LaTeX author packages for each AMS publication. Author packages include instructions for preparing electronic manuscripts, the *AMS Author Handbook*, samples, and a style file that generates the particular design specifications of that publication series. Though $\mathcal{A}_{\mathcal{M}}\mathcal{S}$-LaTeX is the highly preferred format of TeX, author packages are also available in $\mathcal{A}_{\mathcal{M}}\mathcal{S}$-TeX.

Authors may retrieve an author package from e-MATH starting from `www.ams.org/tex/` or via FTP to `ftp.ams.org` (login as `anonymous`, enter username as password, and type `cd pub/author-info`). The *AMS Author Handbook* and the *Instruction Manual* are available in PDF format following the author packages link from `www.ams.org/tex/`. The author package can be obtained free of charge by sending email to `pub@ams.org` (Internet) or from the Publication Division, American Mathematical Society, 201 Charles St., Providence, RI 02904, USA. When requesting an author package, please specify $\mathcal{AMS}$-LaTeX or $\mathcal{AMS}$-TeX, Macintosh or IBM (3.5) format, and the publication in which your paper will appear. Please be sure to include your complete mailing address.

Sending electronic files. After acceptance, the source file(s) should be sent to the Providence office (this includes any TeX source file, any graphics files, and the DVI or PostScript file).

Before sending the source file, be sure you have proofread your paper carefully. The files you send must be the EXACT files used to generate the proof copy that was accepted for publication. For all publications, authors are required to send a printed copy of their paper, which exactly matches the copy approved for publication, along with any graphics that will appear in the paper.

TeX files may be submitted by email, FTP, or on diskette. The DVI file(s) and PostScript files should be submitted only by FTP or on diskette unless they are encoded properly to submit through email. (DVI files are binary and PostScript files tend to be very large.)

Electronically prepared manuscripts can be sent via email to `pub-submit@ams.org` (Internet). The subject line of the message should include the publication code to identify it as a Memoir. TeX source files, DVI files, and PostScript files can be transferred over the Internet by FTP to the Internet node `e-math.ams.org` (130.44.1.100).

Electronic graphics. Comprehensive instructions on preparing graphics are available at `www.ams.org/jourhtml/graphics.html`. A few of the major requirements are given here.

Submit files for graphics as EPS (Encapsulated PostScript) files. This includes graphics originated via a graphics application as well as scanned photographs or other computer-generated images. If this is not possible, TIFF files are acceptable as long as they can be opened in Adobe Photoshop or Illustrator. No matter what method was used to produce the graphic, it is necessary to provide a paper copy to the AMS.

Authors using graphics packages for the creation of electronic art should also avoid the use of any lines thinner than 0.5 points in width. Many graphics packages allow the user to specify a "hairline" for a very thin line. Hairlines often look acceptable when proofed on a typical laser printer. However, when produced on a high-resolution laser imagesetter, hairlines become nearly invisible and will be lost entirely in the final printing process.

Screens should be set to values between 15% and 85%. Screens which fall outside of this range are too light or too dark to print correctly. Variations of screens within a graphic should be no less than 10%.

Inquiries. Any inquiries concerning a paper that has been accepted for publication should be sent directly to the Electronic Prepress Department, American Mathematical Society, 201 Charles St., Providence, RI 02904, USA.

Editors

This journal is designed particularly for long research papers, normally at least 80 pages in length, and groups of cognate papers in pure and applied mathematics. Papers intended for publication in the *Memoirs* should be addressed to one of the following editors. In principle the Memoirs welcomes electronic submissions, and some of the editors, those whose names appear below with an asterisk (*), have indicated that they prefer them. However, editors reserve the right to request hard copies after papers have been submitted electronically. Authors are advised to make preliminary email inquiries to editors about whether they are likely to be able to handle submissions in a particular electronic form.

***Algebra** to ROBERT GURALNICK, Department of Mathematics, University of Southern California, Los Angeles, CA 90089-1113; email: `guralnic@math.usc.edu`

Algebraic geometry to DAN ABRAMOVICH, Department of Mathematics, Boston University, 111 Cummington St., Boston, MA 02215; email: `abramovic@bu.edu`

***Algebraic number theory** to V. KUMAR MURTY, Department of Mathematics, University of Toronto, 100 St. George Street, Toronto, ON M5S 1A1, Canada; email: `murty@math.toronto.edu`

Algebraic topology and cohomology of groups to STEWART PRIDDY, Department of Mathematics, Northwestern University, 2033 Sheridan Road, Evanston, IL 60208-2730; email: `priddy@math.nwu.edu`

Combinatorics and Lie theory to SERGEY FOMIN, Department of Mathematics, University of Michigan, Ann Arbor, Michigan 48109-1109; email: `fomin@umich.edu`

Complex analysis and complex geometry to DUONG H. PHONG, Department of Mathematics, Columbia University, 2990 Broadway, New York, NY 10027-0029; email: `phong@math.columbia.edu`

***Differential geometry and global analysis** to LISA C. JEFFREY, Department of Mathematics, University of Toronto, 100 St. George St., Toronto, ON Canada M5S 3G3; email: `jeffrey@math.toronto.edu`

Dynamical systems and ergodic theory to ROBERT F. WILLIAMS, Department of Mathematics, University of Texas, Austin, Texas 78712-1082; email: `bob@math.utexas.edu`

***Functional analysis and operator algebras** to MARIUS DADARLAT, Department of Mathematics, Purdue University, 150 N. University St., West Lafayette, IN 47907-2067; email: `mdd@math.purdue.edu`

***Geometric analysis** to TOBIAS COLDING, Courant Institute, New York University, 251 Mercer St., New York, NY 10012; email: `colding@cims.nyu.edu`

***Geometric analysis** to MLADEN BESTVINA, Department of Mathematics, University of Utah, 155 South 1400 East, JWB 233, Salt Lake City, Utah 84112-0090; email: `bestvina@math.utah.edu`

Harmonic analysis to ALEXANDER NAGEL, Department of Mathematics, University of Wisconsin, 480 Lincoln Drive, Madison, WI 53706-1313; email: `nagel@math.wisc.edu`

Harmonic analysis, representation theory, and Lie theory to ROBERT J. STANTON, Department of Mathematics, The Ohio State University, 231 West 18th Avenue, Columbus, OH 43210-1174; email: `stanton@math.ohio-state.edu`

***Logic** to STEFFEN LEMPP, Department of Mathematics, University of Wisconsin, 480 Lincoln Drive, Madison, Wisconsin 53706-1388; email: `lempp@math.wisc.edu`

Number theory to HAROLD G. DIAMOND, Department of Mathematics, University of Illinois, 1409 W. Green St., Urbana, IL 61801-2917; email: `diamond@math.uiuc.edu`

***Ordinary differential equations, and applied mathematics** to PETER W. BATES, Department of Mathematics, Michigan State University, East Lansing, MI 48824-1027; email: `peter@math.msu.edu`

***Partial differential equations** to PATRICIA E. BAUMAN, Department of Mathematics, Purdue University, West Lafayette, IN 47907-1395; email: `bauman@math.purdue.edu`

***Probability and statistics** to KRZYSZTOF BURDZY, Department of Mathematics, University of Washington, Box 354350, Seattle, Washington 98195-4350; email: `burdzy@math.washington.edu`

***Real analysis and partial differential equations** to DANIEL TATARU, Department of Mathematics, University of California, Berkeley, Berkeley, CA 94720; email: `tataru@ math.berkeley.edu`

All other communications to the editors should be addressed to the Managing Editor, WILLIAM BECKNER, Department of Mathematics, University of Texas, Austin, TX 78712-1082; email: `beckner@math.utexas.edu`.

Titles in This Series

795 **Adam Nyman,** Points on quantum projectivizations, 2004

794 **Kevin K. Ferland and L. Gaunce Lewis, Jr.,** The $RO(G)$-graded equivariant ordinary homology of G-cell complexes with even-dimensional cells for $G = \mathbb{Z}/p$, 2004

793 **Jindřich Zapletal,** Descriptive set theory and definable forcing, 2004

792 **Inmaculada Baldomá and Ernest Fontich,** Exponentially small splitting of invariant manifolds of parabolic points, 2004

791 **Eva A. Gallardo-Gutiérrez and Alfonso Montes-Rodríguez,** The role of the spectrum in the cyclic behavior of composition operators, 2004

790 **Thierry Lévy,** Yang-Mills measure on compact surfaces, 2003

789 **Helge Glöckner,** Positive definite functions on infinite-dimensional convex cones, 2003

788 **Robert Denk, Matthias Hieber, and Jan Prüss,** $\mathcal{R}$-boundedness, Fourier multipliers and problems of elliptic and parabolic type, 2003

787 **Michael Cwikel, Per G. Nilsson, and Gideon Schechtman,** Interpolation of weighted Banach lattices/A characterization of relatively decomposable Banach lattices, 2003

786 **Arnd Scheel,** Radially symmetric patterns of reaction-diffusion systems, 2003

785 **R. R. Bruner and J. P. C. Greenlees,** The connective K-theory of finite groups, 2003

784 **Desmond Sheiham,** Invariants of boundary link cobordism, 2003

783 **Ethan Akin, Mike Hurley, and Judy A. Kennedy,** Dynamics of topologically generic homeomorphisms, 2003

782 **Masaaki Furusawa and Joseph A. Shalika,** On central critical values of the degree four L-functions for GSp(4): The Fundamental Lemma, 2003

781 **Marcin Bownik,** Anisotropic Hardy spaces and wavelets, 2003

780 **S. Marmi and D. Sauzin,** Quasianalytic monogenic solutions of a cohomological equation, 2003

779 **Hansjörg Geiges,** h-principles and flexibility in geometry, 2003

778 **David B. Massey,** Numerical control over complex analytic singularities, 2003

777 **Robert Lauter,** Pseudodifferential analysis on conformally compact spaces, 2003

776 **U. Haagerup, H. P. Rosenthal, and F. A. Sukochev,** Banach embedding properties of non-commutative L^p-spaces, 2003

775 **P. Lochak, J.-P. Marco, and D. Sauzin,** On the splitting of invariant manifolds in multidimensional near-integrable Hamiltonian systems, 2003

774 **Kai A. Behrend,** Derived ℓ-adic categories for algebraic stacks, 2003

773 **Robert M. Guralnick, Peter Müller, and Jan Saxl,** The rational function analogue of a question of Schur and exceptionality of permutation representations, 2003

772 **Katrina Barron,** The moduli space of $N = 1$ superspheres with tubes and the sewing operation, 2003

771 **Shigenori Matsumoto,** Affine flows on 3-manifolds, 2003

770 **W. N. Everitt and L. Markus,** Elliptic partial differential operators and symplectic algebra, 2003

769 **Jie Wu,** Homotopy theory of the suspensions of the projective plane, 2003

768 **R. Höpfner and E. Löcherbach,** Limit theorems for null recurrent Markov processes, 2003

767 **Po Hu,** S-modules in the category of schemes, 2003

766 **Su Gao and Alexander S. Kechris,** On the classification of Polish metric spaces up to isometry, 2003

765 **Robert Bieri and Ross Geoghegan,** Connectivity properties of group actions on non-positively curved spaces, 2003

764 **J. Spandaw,** Noether-Lefschetz problems for degeneracy loci, 2003

763 **Yasuyuki Kachi and Eiichi Sato,** Segre's reflexivity and an inductive characterization os hyperquadrics, 2002

762 **Leiba Rodman, Ilya M. Spitkovsky, and Hugo Woerdeman,** Abstract band method via factorization, positive and band extensions of multivariable almost periodic matrix functions, and spectral estimation, 2002

761 **Oliver Druet and Emmanuel Hebey,** The AB program in geometric analysis : Sharp Sobolev inequalities and related problems, 2002

760 **Markus Banagl,** Extending intersection homology type invarients to non-Witt spaces, 2002

759 **Donald M. Davis,** From representation theory to homotopy groups, 2002

758 **Alan Forrest, John Hunton, and Johannes Kellendonk,** Topological invariants for projection method patterns, 2002

757 **Douglas Bowman,** q-difference operators, orthogonal polynomials, and symmetric expansions, 2002

756 **José Ignacio Cogolludo-Agustín,** Topological invariants of the complement to arrangements of rational plane curves, 2002

755 **M. A. Mandell and J. P. May,** Equivariant orthogonal spectra and S-modules, 2002

754 **Edward L. Green, Idun Reiten, and Øyvind Solberg,** Dualities on generalized Koszul algebras, 2002

753 **Daniel Panazzolo,** Desingularization of nilpotent singularities in families of planar vector fields, 2002

752 **Linus Kramer,** Homogeneous spaces, Tits buildings, and isoparametric hypersurfaces, 2002

751 **Bruce Allison, Georgia Benkart, and Yun Gao,** Lie algebras graded by the root systems BC_r, $r \geq 2$, 2002

750 **Masaki Izumi and Hideki Kosaki,** Kac algebras arising from composition of subfactors: General theory and classification, 2002

749 **Nanhua Xi,** The based ring of two-sided cells of affine Weyl groups of type $\widetilde{A}_{n-1}$, 2002

748 **Jürgen Ritter and Alfred Weiss,** The lifted root number conjecture and Iwasawa theory, 2002

747 **Armand Borel, Robert Friedman, and John W. Morgan,** Almost commuting elements in compact Lie groups, 2002

746 **Peter Niemann,** Some generalized Kac-Moody algebras with known root multiplicities, 2002

745 **Mikhail A. Lifshits and Werner Linde,** Approximation and entropy numbers of Volterra operators with application to Brownian motion, 2002

744 **Roger Chalkley,** Basic global relative invariants for homogeneous linear differential equations, 2002

743 **Heng Sun,** Spectral decomposition of a covering of $GL(r)$: the Borel case, 2002

742 **J. E. Gilbert, Y. S. Han, J. A. Hogan, J. D. Lakey, D. Weiland, and G. Weiss,** Smooth molecular functions and singular integral operators, 2002

741 **Francisco Santos,** Triangulations of oriented matroids, 2002

740 **Rick Durrett,** Mutual invadability implies coexistence in spatial models, 2002

739 **Georgios K. Alexopoulos,** Sub-Laplacians with drift on Lie groups of polynomial volume growth, 2002